The Biology
of
Symbiotic Fungi

The Biology
of
Symbiotic Fungi

Roderic Cooke

Botany Department
University of Sheffield

JOHN WILEY & SONS

London · New York · Sydney · Toronto

Library of Congress Cataloging in Publication Data:

Cooke, Roderic.
    The Biology of symbiotic fungi.

    Includes bibliographies and index.
    1. Fungi. 2. Symbiosis. I. Title.
QK604.C63   589'.2'04524   76-56175

ISBN 0 471 99467 7

Typeset in IBM Baskerville by Preface Ltd, Salisbury, Wilts
Printed by The Pitman Press, Bath, Avon.

Preface

The appearance of increasing numbers of symposia, review articles and research papers concerned with symbiology attests to a steadily increasing interest in both theoretical and experimental aspects of this broad subject. This growth in interest is not confined to research scientists, courses concerned with interactions among organisms are becoming an increasingly important part of university teaching curricula in biological sciences. Many interesting, and ecologically or economically important, symbioses involve fungi, and aspiring students of these associations are relatively well served by an extensive literature. This literature is, however, commonly either extremely specialized within particular aspects of fungal symbiosis or is so diffuse that much effort must be expended to obtain very little easily assimilable information. There is a lack of texts to which students can recourse for concise accounts of more than a few aspects of this subject. This book has been written for undergraduate and postgraduate students who have some knowledge of mycology and it attempts to outline all major symbioses that exist between fungi and either animals or plants. Emphasis is placed on the manner in which these various associations function or, where details are not fully known, how they might possibly function.

A major difficulty has been to make a decision as to the limits of the term 'symbiosis'. After a great deal of vacillation the position finally adopted is that of acceptance of de Bary's original broad concept of symbiosis as meaning simply 'living together'. Some may, quite understandably, demur and consider that the approach taken here is much too broad. However, there is a current and growing concern with the exploration of this wider view, and with the placing of apparently disparate symbiotic associations in a common context. This allows at least a beginning to be made on a synthetic treatment of symbiology as a whole.

Another difficulty has been the fact that symbioses between fungi and plants have received, and still are receiving, intensive study, while those involving associations between fungi and animals have always been relatively neglected. This means that while it has been possible to discuss fungus–plant symbioses, for the most part, in terms of established physical and physiological phenomena this has not always been possible for fungus–animal symbioses. The great disparity between our knowledge of fungus–plant and fungus–animal associations, together with the many differences in their essential nature, also precludes a comprehensive comparative approach, at least for the present. This book therefore falls into two distinct parts, the first concerned with animals the second with plants. Rather more detail is given for fungus–animal symbioses since these, in general, are the least familiar, and the literature concerning them is frequently not easily available.

In a book of this length it has been necessary to make a number of omissions, particularly in those sections concerned with fungi that cause diseases in animals or plants. For example, fungi of molluscs, crustacea, fish, reptiles and birds receive only passing mention. Although these fungi are by no means unimportant, knowledge of their biology is so fragmentary that any kind of treatment other than a catalogue of names and symptoms is not feasible. With respect to plants, details of their initial infection by fungi are not dealt with and there is little mention of either mechanisms of host resistance or the production and mode of action of fungal toxins. All these topics have received more than adequate treatment in a wide range of good standard textbooks. Neutral fungi of the rhizosphere and phylloplane are also not discussed since so little is yet known of their status and biology.

It is important to remember that this is a book about fungi and not, primarily, about their animal or plant associates. For a number of reasons fungus–animal symbioses have long been the province of entomologists, microbiologists, medical scientists and other non-mycological specialists. It has sometimes been necessary to reinterpret their conclusions in mycological terms.

It seems inevitable that anything written about symbiology, particularly its more fundamental aspects, should provoke some degree of controversy and it seems unlikely that this book will be unique in escaping comment and critism. It aims to do three things. First, it seeks to erect a simple conceptual framework within which students may study fungal symbioses with a minimum of confusion. Second, it tries to present a suitable body of information that will allow students to grasp the basic mechanisms involved in various kinds of symbioses. Finally, it attempts to leaven this information with a modest amount of speculation. In Charles Dickens's novel *Hard Times*, Mr. Gradgrind asserts that '*Facts alone are wanted in life. Plant nothing else, and root out everything else. You can only form the minds of reasoning animals upon Facts. Nothing else will ever be of service to them.*' While not disagreeing entirely with this philosophy this book does try to go a little further than the facts.

July 1976   Roderic Cooke
Sheffield

Acknowledgements

I would like to thank Jill Gray and Janet Lambden, who efficiently typed the manuscript from my indecipherable handwriting, and Glyn Woods for all the photographic work. I am also extremely grateful to Elizabeth Craghill, who prepared the great majority of the figures, and to Margaret Cooke who helped to correct the final typescript and proofs. I also thank those authors and other copyright holders, who are all cited in the appropriate place, for their permission to use illustrative material.

Contents

Concepts of Fungal Symbiosis

Chapter 1

Classification of Symbiotic Fungi

Associations exist between fungi and a wide range of both autotrophic and heterotrophic organisms. The nature of them is diverse, there being found varying degrees of intimacy, permanence and nutritional interdependence. Attempts have been made to produce generally acceptable classifications of the associations, and some valuable conceptual analysis of them have been made (Harley, 1968; Lewis, 1973, 1974; Luttrell, 1974). Despite these analyses and efforts to impose precision, the terminology used to refer to associations between fungi and other organisms is confused and is likely to remain so for some time to come. Some of the problems of terminology arise from purely semantic difficulties but the most important spring from genuinely irreconcilable points of view. Although it is manifestly difficult to propose a universally acceptable scheme of classification, some of the basic virtues that it might ideally embody are relatively easy to visualize. First, the terminology used should be free of ambiguity, be of a kind likely to gain value in a general biological context, and be aimed at reducing the current proliferation of terms. Second, the scheme should be succinct and indicate clearly the degree of dependence, and the fundamental bases for the dependence, of those fungi contained within it. Finally, it should take account of, be related to, or fit within, classifications of associations that involve other kinds of organisms apart from fungi. What follows is the outline of a system which attempts to satisfy these criteria. It should be emphasized that it is meant to be neither rigid nor final, but aims only to provide a framework upon which a treatment of the nature of fungal associations may be built.

Fungi are chemoheterotrophic, depending on organic compounds as their principal energy and carbon sources. They must derive these compounds from either non-living organic materials or from living tissues. In the latter situation the fungi come into association with a suitable host organism and this relationship with their hosts, however temporary, is that of a common life. The word *symbiosis* was originally used strictly to describe this concept of a common life, although it was at the same time recognized that every kind of gradation could be found in the nature of fungus—host interactions (de Bary, 1887).

Gradually, within mycology, the use of the term symbiosis has come to be erroneously restricted to associations that are characterized by *mutualism*, that is where fungus and host either wholly or partly maintain one another. Currently there does, however, seem to be movement towards a restoration of its initial broad meaning, although even so there is some reluctance to apply it to associations that do not involve prolonged or permanent intimate contact between fungus and host (Lewis, 1973; 1974). This reluctance is difficult to understand, but may be related to the fact that, within mycology, concepts of symbiosis have been based principally on relationships between fungi and autotrophic plants, where prolonged and permanent intimate contacts are common. If, however, relationships between fungi and animals are considered there are clear examples of what are undeniably symbioses in which fungus and host are closely associated, but where there is no intimate contact between their somatic tissues, although at some time during its life cycle the fungus may be held in a living condition within specialized host organs (Francke-Grosmann, 1967; Hartzell, 1967; Weber, 1972). Symbiosis is, therefore, used here in the all-embracing sense to refer to *all* associations where fungi come into contact with a living host from which they obtain, in a variety of ways, either major or minor metabolites or nutrients.

The most obviously important characteristic of a symbiosis is whether harm or benefit results from the association, and with respect to this there are three possibilities for fungal symbiosis. First, the fungus may be *antagonistic* towards its partner, and cause it slight or severe harm either directly or indirectly. Antagonistic symbionts are normally referred to as *parasites* or, if they cause damage sufficiently severe to produce easily recognized disease symptoms in their host, as *pathogens*. Antagonist and antagonistic are perhaps less ambiguous terms than parasite and parasitic and may eventually come to replace the latter, although this is unlikely to take place rapidly. Second, the fungus may be *mutualistic* towards its partner which, as a result, benefits in some way from the association. Between antagonism and mutualism a third possibility, that of *neutralism*, may be distinguished. A fungus may be considered to be a neutral symbiont if it is consistently found associated with a host upon which it is absolutely dependent but on which it has no obvious deleterious or beneficial effect. The dependence on the part of the neutral fungus may be for either a long-term association or a transient one. Fungi that require a long-term association cannot usually live in nature when separated from their host. Where contact is transient, then the fungi usually require passage through or a brief housing within their host, which is usually an animal, to complete a mainly free-living life cycle. Neutralism corresponds closely with the term *commensalism* where the latter is used to describe a situation in which only one partner profits from an association. Neutral fungi have, in certain situations, been referred to as *harmless parasites*, a confusing hybrid since harm is implicit in the term parasite.

The second important characteristic of a fungal symbiosis is the degree of dependence of the fungus upon association with its partner, whether the relationship is necessary, that is *obligate*, or contingent, that is *facultative*. The terms obligate and facultative are mutually exclusive and are here applied strictly in the ecological sense, referring to the situation obtaining under natural conditions. Fungi that are obligate symbionts have no capacity for a free-living existence, other than as propagules, in the absence of a suitable host. In contrast, facultative symbionts, while being always potential symbionts if suitable ecological conditions arise, do have a well-developed free-living capability. In the past, particularly with respect to fungi that are antagonistic towards higher plants, the word obligate has been used in a rather different context. Where it was apparently not possible to bring an antagonistic fungus into axenic culture the term *obligate parasite* was applied to it, the implication being that it was absolutely nutritionally dependent upon its host for vegetative growth. Since it is becoming increasingly clear that a large number, if not all, of these hitherto unculturable fungi will eventually be grown in axenic culture, they can no longer be considered to be obligately dependent upon their hosts in that particular narrow sense.

Fungal symbionts also have certain nutritional characteristics that are determined by the source and derivation of their nutrients. Three modes of nutrition are possible. First, a fungus may be *saprotrophic*, that is derive organic compounds directly from the non-living components of its immediate environment. Second, it may be *necrotrophic*, deriving its organic nutrients from the dead cells of organisms which it has itself killed. Necrotrophs are in the end behaving in a manner similar to saprotrophs but are continually creating their own dead organic substrates throughout the duration of the symbiosis. Finally, a fungal symbiont may be *biotrophic* and be capable of deriving its organic nutrients only from the living cells of its host. Should the host or that part of it occupied by the fungus for some reason die, then the fungus also either dies, or is forced into a period of dormancy or inactivity, in the form of a resting mycelium or propagules, until a new association can be initiated. Unlike the situation obtaining with respect to the degree of dependence on symbiotic association, where a fungus must be either ecologically obligately symbiotic or ecologically facultatively symbiotic, these three modes of nutrition are not mutually exclusive. A single fungus may show two, or even all three, kinds of nutrition during its life cycle, and changes in mode are frequently caused by changes in ecological conditions.

Biotrophs commonly show highly developed physiological specialization combined with a relatively restricted host range. Both these characteristics indicate that biotrophs have suffered loss of adaptability while evolving to their present degree of fitness to exploit particular ecological niches. Some of this loss of adaptability may be expressed in strictly nutritional terms, that is a fungus may have developed an absolute requirement for complex organic nutrients that it cannot itself synthesize and which it must obtain from its host. However, other aspects of the loss may be reflected in a diminution in other kinds of versatility, namely a lack of those characters, both physical and physiological, which allow a fungus to compete for space and nutrients with other micro-organisms. Similarly, some necrotrophic fungi have very simple nutrient requirements which can be satisfied in axenic culture, but they cannot compete saprotrophically in natural situations with other micro-organisms. They are thus often obligately dependent on a symbiotic association that allows them to exploit host tissues necrotrophically and so enables them to occupy and command their own dead organic substrates. Obligately symbiotic fungi which are entirely saprotrophic during symbiosis resemble these necrotrophs in being unable to compete, and so they usually lack a free-living saprotrophic phase in their life cycle. They succeed as saprotrophs because they occupy peculiar ecological niches that free-living saprotrophs are unable to invade or exploit. Many fungi, therefore, are not ecologically obligately symbiotic solely for reasons related directly to extremely specialized nutrient requirements, but they are so because of a number of more complex factors.

Using the criteria of mutualism, neutralism and antagonism, combined with those of obligate or facultative dependency, it is possible to divide symbiotic fungi into six biological groups (Table 1). It should be remembered that these groups are to some extent fluid in that some of the fungi within them can, at different times, move from one group to another, depending on the stage at which they happen to be in their life cycle. While a fungus can only be either obligately or facultatively dependent, and cannot shift from one category to another, it is possible for some fungi to show neutralism, mutualism and antagonism towards their hosts during different phases of their interaction. It is also possible for a fungus to have mutualistic and antagonsitic effects on its host simultaneously and over a relatively long period of time. However, whether there are phased changes in, or simultaneous mixtures of, effects there is usually a very clear overall outcome for the host, and this normally allows such fungi to be placed reasonably firmly in one of the groups.

In this scheme modes of nutrition are not used to further divide the six groups. Such subdivision is not possible in all groups and where it is possible it would lead only to confusion. This is mainly due to the fact, remarked previously, that some fungi can show all three nutritional modes in a single host depending on the location of the fungus, the physiological state of the host, or prevailing ecological conditions. In

Table 1
Symbiotic fungi divided into biological groups according to their
degree of dependence and effect on their hosts

	Degree of dependence	
	Facultatively symbiotic	Obligately symbiotic
Antagonistic	Group 1	Group 2
Neutral	Group 3	Group 4
Mutualistic	Group 5	Group 6

Effect on host

some cases it is also impossible to draw a clear line between saprotrophy and necrotrophy or between necrotrophy and biotrophy.

Group 1. Antagonistic Facultative Symbionts

These are fungi that have a moderate or well-developed ability for a free-living existence and which are also potentially symbiotic. When free-living they are saprotrophic but when symbiotic they are necrotrophic.

Those that become associated with plants normally produce diseases that are either characterized by extensive breakdown of living host tissues, for example damping off, soft rots and related diseases, or by fatal wilting of the host due to growth of the fungus in its water-conducting elements.

Those that become symbiotic with animals, including man, cause mild, debilitating or fatal diseases which are characterized by restricted or extensive destruction of living host tissues. In addition, some fungi have the ability to capture, kill and subsequently utilize microscopic animals, for example protozoa, rotifers, nematode worms and mites. This characteristic of predation allows a transient necrotrophic symbiosis to be established with a mobile host.

Finally, several examples are known of fungi which live necrotrophically on other fungi.

Group 2. Antagonistic Obligate Symbionts

This group comprises fungi that entirely lack or have only a poorly-developed capacity for a free-living existence in natural conditions and require a symbiotic association for normal phenotypic development. When symbiotic they are either necrotrophic or biotrophic.

The majority of such fungi that become associated with plants are biotrophic, producing diseases of either above-ground or subterranean organs. These diseases are usually typified by a number of characteristic symptoms. For example, there is retention of chlorophyll and accumulation of host metabolites around infection sites, the development of hyperauxiny and exaggerated growth responses, and there are nuclear and nucleolar changes within host cells affected by the fungus. Typical diseases of this kind are rusts, smuts, powdery and downy mildews, ergot and some root and tuber diseases. There are fungi in this group, however, that are not biotrophic and which do have some capacity, albeit restricted, for a free-living saprotrophic existence under suitable conditions. Such fungi can survive in a mycelial form for considerable periods in dead, infected host tissues and can use these tissues as a food base from which to grow and eventually contact suitable living hosts. Having achieved this their nutrition then becomes necrotrophic. Typical examples are fungi that cause root or butt rots of forest trees. There is some evidence that many fungi that cause rotting, damping-off and wilting, and which have hitherto been considered as facultative symbionts, should be placed in this group. Finally, a few species, typically leaf spotting fungi, are at first biotrophic upon their hosts but later, when infected leaf tissue dies as a result of their activity, they become necrotrophic and, finally, saprotrophic. Areas of dead tissue are, however, restricted in exent and the fungi are confined to these areas.

Antagonistic obligate symbionts that come into association with animals are for the most part necrotrophic. Some resemble those of Group 1 in that they can cause diseases of varying severity or can capture, kill and consume microscopic animals, principally protozoa. Some fungi that exploit microscopic animals are entirely endosymbiotic and development outside the host is confined to the production of spore-producing organs. Some endosymbiotic fungi, almost exclusively species associated with protozoa, are biotrophs.

Also included in this group are fungi that inhabit the skin and other keratinized parts of higher animals, where they can produce a group of diseases collectively termed dermatophytoses. It is not altogether clear whether they are living simply as saprotrophs on dead, keratinized tissues or whether their activities contribute to the extra production of such tissues. In the latter case they could be considered to be necrotrophs.

Finally, a number of fungi are known which are biotrophic upon other fungi.

Group 3. Neutral Facultative Symbionts

These are fungi that live upon or within a host on which they have no obvious beneficial or deleterious effect. Their mode of nutrition is saprotrophic and they derive their nutrients from either the dead, organic parts of their hosts or from diffusates originating from their hosts. These fungi do have a capacity for a saprotrophic existence in the absence of suitable hosts although they may not always be commonly found away from them. Sometimes, if there is a shift in the physiological status of the host, a Group 3 species may become weakly antagonistic and necrotrophic, so that it then has the properties of a Group 1 organism. Examples of Group 3 fungi are found among species that comprise the mycoflora of the phylloplane, rhizoplane and rhizosphere of plants.

Group 4. Neutral Obligate Symbionts

Obligate neutrals resemble facultative neutrals of Group 3 in that they inhabit hosts on which they have no obvious effect. However, unlike members of the latter group, they are obviously dependent on either a long-term or transient association with their hosts for normal phenotypic development.

Biotrophic fungi that have been described as 'harmless parasites' or 'harmless endophytes' of plant roots are probably obligately neutral, as are probably many saprotrophic fungi that are characteristic of the phylloplane, rhizoplane and rhizosphere. It should be emphasized that some endophytic fungi fall within this group only because of lack of experimental evidence as to their possible antagonistic or mutualistic effect on their hosts. In many cases the effects of infection might only occur at the cellular level, and then only upon the host cell that is actually occupied by fungus. Since the proportion of occupied cells is often relatively small, no marked changes in the gross physiology of the host would be likely to occur. In addition, the free-living capability of many endophytes is completely unknown.

Obligate neutrals are also associated with animals, particularly Arthropods, where they grow for the most part saprotrophically either on the integument or within the gut. There is no evidence that these fungi have a capacity for a free-living life form and their association with their hosts appears to be a permanent one.

A number of neutral fungi are associated with warm-blooded animals and in these cases the symbiosis can be either long-term or transient. Long-term associations involve saprotrophic fungi that inhabit the skin and its appendages or the mucous membranes. Like Group 3 fungi, they may become antagonistic and necrotrophic if there is a shift in the physiological status of their host which favours this mode of life. Transient symbioses involve a heterogeneous group of 'passage fungi' that have a well-developed capacity for a free-living saprotrophic existence but which, nevertheless, require a brief symbiosis, usually while in the form of spores, to successfully complete their life cycle. The outstanding examples are coprophilous fungi whose spores are either completely or partially dormant. Passage through the gut of a host, usually but not invariably a warm-blooded animal, alleviates this dormancy so that germination subsequently occurs in the voided faeces and phenotypic development then proceeds.

Group 5. Mutualistic Facultative Symbionts

These symbionts have some capacity for a free-living existence but are normally found in association with other organisms. When free-living their nutrition is saprotrophic but during symbiosis it is biotrophic. Examples of such fungi are those species which are facultatively mycorrhizal with the roots of higher plants, and species that are capable of forming lichen-like associations with algae.

For reasons that are not at all clear, there appear to be no known mutualistic fungi that are facultatively symbiotic with animals, except perhaps for a few species which may act as auxiliary food fungi for wood-boring beetles.

Group 6. Mutualistic Obligate Symbionts

Fungi in this group have either no capacity or a poor capacity for free-living saprotrophy in nature. Those associated with plants normally exist as biotrophs in mycorrhizal associations or in a lichenized form.

Those associated with animals can either be saprotrophic or biotrophic. Saprotrophic symbionts either act directly as food fungi for insects or modify a normally non-utilizable organic food source, for example wood, so that it can be eaten by insects. During certain periods in the life of the insect the fungi may be housed within specialized host organs within which their nutrition remains saprotrophic, host secretions being utilized as a nutrient source. Although these fungi have an ability to grow saprotrophically outside the insect, they can only do so successfully in certain restricted ecological niches into which they are introduced by their insect hosts. There is total dependence on their hosts for both establishment and dissemination.

Biotrophic species are also commonly associated with insects. In their most highly-developed form they are intracellular within the cytoplasm of specialized host cells, the latter being part of distinct, specialized fungus-containing organs, and they are apparently incapable of an active life outside these.

Several points may be made concerning this scheme of symbiotic groupings. There is notable heterogeneity among the fungi in any single group, there being fungi with widely differing taxonomic and physiological characteristics within the same group. Similarly, fungi with close taxonomic and physiological affinities are located in different groups.

In addition, symbiotic fungi are known which are very closely related to non-symbiotic saprotrophs that lie outside these groupings. The groups generally have a degree of plasticity, to the extent that some of their constituent fungi can be moved from group to group depending on those ecological or physiological circumstances obtaining at any particular time.

Within each group fungi that are symbiotic with plants are placed with their counterparts that are symbiotic with animals (and sometimes symbiotic with other fungi), yet the fundamental nature of these symbioses can be very different. Furthermore, if the distribution of members of the taxonomic subdivisions of fungi among the six symbiotic groups is examined in relation to the kind of host involved, then within each subdivision fewer combinations of associations are found between fungi and animals than between fungi and plants (Table 2). This might, of course, only reflect a lack of knowledge of animal-symbiotic fungi, but it does seem that in absolute terms examples of associations between fungi and plants are much more common and widespread than those between fungi and animals. It is striking that, while biotrophy is frequent among plant-symbiotic fungi it is relatively uncommon in animal-symbiotic fungi. Although Table 2 shows the spread of symbiotic relationships so far known for the major fungal subdivisions, it should be noted that in some symbiotic groups a

Table 2

Major taxonomic divisions of fungi and known symbiotic associations (o) between their members and plants or animals

Nature of host	Group 1 Facultative Antagonistic Plant	Animal	Group 2 Obligate Antagonistic Plant	Animal	Group 3 Facultative Neutral Plant	Animal	Group 4 Obligate Neutral Plant	Animal	Group 5 Facultative Mutualistic Plant	Animal	Group 6 Obligate Mutualistic Plant	Animal
Myxomycota			o									
Mastigomycotina	o	o	o	o			o					
Zygomycotina	o	o		o				o			o	
Ascomycotina	o	o	o	o	o		o	o	o		o	o
Basidiomycotina	o	o	o	o	o			o	o		o	o
Deuteromycotina	o	o	o	o	o		o		o		o	o

Taxonomic subdivisions of fungi

12

particular subdivision may occasionally be represented by only a single order, family, genus, or even species, of fungus. However, this situation also may be due to lack of information on fungal symbioses in general.

These problems present difficulties that in the main preclude a concise synthetic treatment of fungal symbioses. It is certainly at present impossible to treat plant and animal symbioses in a comparative way, since experimental data are largely not available for the latter. For this reason, while the six symbiotic groups for the moment act as a useful framework within which to set symbiotic fungi, to treat and describe symbioses simply group by group would lead to confusion. In the text which follows, fungi symbiotic with plants or animals are, therefore, dealt with separately, and within their separate treatments antagonists, mutualists and neutrals are each considered in turn. With respect to neutrals, treatment is restricted to those fungi which are symbiotic with animals.

Bibliography

De Bary, A., 1887. *Comparative Morphology and Biology of the Fungi, Mycetozoa and Bacteria,* Oxford.
Francke-Grosmann, H., 1967. In S. M. Henry (ed.), *Symbiosis,* Vol. 2, p. 141–205, Academic Press, New York.
Harley, J. L., 1968. *Trans. Br. mycol. Soc.,* 51:1–12.
Hartzell, A., 1967. In S. M. Henry (ed.), *Symbiosis,* Vol. 2, p. 107–140, Academic Press, New York.
Lewis, D. H., 1973. *Biol Rev.,* 48: 261–278.
Lewis, D. H., 1974. In M J. Carlile and J. J. Skehel (eds.), *Evolution in the Microbial World,* p. 367–392. 24th *Symp. Soc. gen. Microbiol.* Cambridge University Press.
Luttrell, E. S., 1974. *Mycologia,* 66: 1–15.
Weber, N. A., 1972. *Gardening Ants: the Attini,* American Philosophical Society, Philadelphia.

Antagonistic Symbioses with Animals

Chapter 2

Fungi that Capture Microscopic Animals

Many fungi that inhabit decomposing organic matter, soil, or fresh water, capture and consume microscopic animals of various kinds. Although taxonomically diverse these fungi form a natural biological group united by their common zoophagous habit, and all have similar morphological adaptations which enable them to pursue this mode of life. They are a remarkable example of parallel evolution. (Drechsler, 1941; Duddington, 1955, 1956, 1957). Their hosts are mobile, frequently moving vigorously, so that the fundamental problems which the fungi have to overcome are to catch, immobilize and efficiently exploit the animals.

Within this group are two distinct series of fungi. First, there are those species that capture relatively small, slow-moving animals, mainly amoebae and other protozoa. Second, there are those that capture larger and quicker moving animals, for example nematodes, rotifers and, occasionally, mites. The first group comprises zygomycetous fungi together with a few Hyphomycetes, while the second group is made up almost entirely of hyphomycetous species but also includes some Zygomycetes. The zygomycetous fungi in both groups belong almost exclusively to a single order, the Zoopagales.

1. Fungi That Capture Protozoa

These species usually capture amoebae, but testaceous rhizopods are also exploited by some. There is a degree of host specificity in that those fungi which capture amoebae do not capture shelled rhizopods and those that capture the latter do not attack amoebae. It has been suggested that, additionally, the fungi may be specific at the host species level, but there is no strong evidence for this.

In the Zoopagales there are four protozoon-capturing genera, *Acaulopage, Cystopage, Stylopage* and *Zoopage*. All have a mycelium of very fine, sparingly-branched hyphae and, with few exceptions, all species capture soil amoebae. An amoeba moving across the substratum comes into contact with a hypha and is held to it by adhesion. The animal then withdraws its pseudopodia and becomes immobile. A fine,

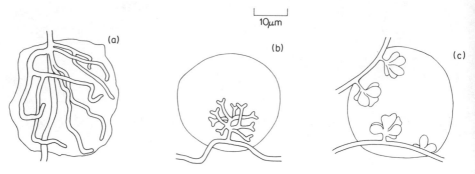

Figure 1   Assimilative hyphae of some Zoopagales within dead amoebae: (a) *Stylopage araea*; (b) *Acaulopage macrospora; (c) Zoopage phanera*. Redrawn from Drechsler, 1935*b*; by permission of *Mycologia*

lateral penetration peg is produced by the fungus at the point of contact between host and hypha and this usually branches within the animal to form an assimilative organ. The protoplast of the host breaks down and, when this process is complete, the cytoplasm of the assimilative organ is withdrawn into the parent hypha.

Occasionally, small droplets of yellow, viscous material are secreted at the point of contact but for the most part hyphae are not very obviously adhesive even when contacted by a potential host. The structure of assimilative organs is constant within a species but inter-specific differences may be very wide. They can either be closely similar to vegetative hyphae or very different from them (Figure 1 a–c). When different they are dichotomously branched, bush-like, or lobed to varying degrees. In some species they are very small in relation to the size of the host and in these cases a host may become penetrated at several points. Without exception, death after capture is very rapid and usually occurs before formation of the assimilative organ is complete. Even when the organ is minute, there is rapid degeneration of the host, and nutrition of these fungi appears to be entirely necrotrophic. Some species are remarkable in that their conidia, as well as their hyphae, are adhesive. If a host is contacted by a spore, the adhering conidium germinates, and its fine germ tube penetrates the pellicle to give rise to an assimilative organ. As the host's body is utilized the conidium produces an external mycelium upon which further hosts may then be caught (Figure 2 a, b).

Three species, *A. crobylospora, Z. toechospora* and *Z. tryphera* capture protozoa other than amoebae (Drechsler, 1973*a*, 1974*a,b*). *A. crobylospora* exploits the proteomyxan organism *Leptomyxa reticulata*. Protoplasmic masses become attached to hyphae and are then penetrated by short, curved branches that act as absorptive organs (Figure 3a). *Z. toechospora* attacks the testate protozoon *Euglypha*

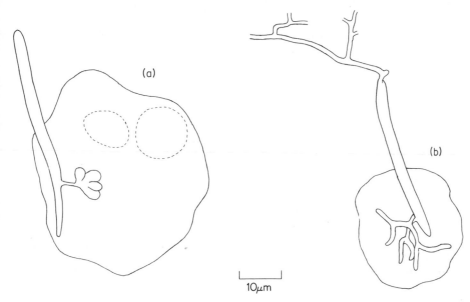

Figure 2   *Stylopage rhabdospora*: (a) adhesive conidium producing an assimilative hypha within an amoeba; (b) subsequent development of the assimilative hypha and production of a mycelium external to the host. Redrawn from Drechsler, 1946; by permission of *Mycologia*

laevis. The animals are caught by the adhesion of the testa to hyphae and an assimilative branch then grows through the opening in the testa within which it finally becomes coiled (Figure 3b). The animal remains alive during at least the early part of this process, and trapped individuals have been observed to lay down platelets within the testa in an attempt to form a barrier against penetration of the fungus to the deep cytoplasm and nucleus. *Z. tryphera* attacks another testaceous rhizopod, *Geococcus vulgaris*, which is then exploited by means of a small dichotomously-branched hypha.

Four Hyphomycetes are known that trap protozoa. *Dactylella tylopaga* captures amoebae while *D. passalopaga, Pedilospora dactylopaga* and *Tridentaria carnivora* attack shelled rhizopods of various kinds (Drechsler, 1934, 1936, 1937b). All except *D. passalopaga* have adhesive hyphae or adhesive hyphal branches (Figure 3c). The hyphae of *D. passalopaga* are not adhesive and yet the fungus can capture *Geococcus vulgaris* and *Euglypha laevis*. Both these animals may attempt to feed upon its hyphae the 'mouth' being applied to the hyphal surface. The zone of contact between the animal and the fungus is then sealed by some kind of secretion from the protozoon which thus becomes firmly attached. The hyphal wall is perforated and fungal cytoplasm is drawn into the animal, but the fungus responds by

18

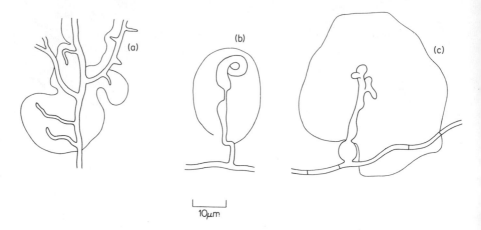

Figure 3   Assimilative hyphae of fungi that capture protozoa: (a) *Acaulopage crobylospora* within the proteomyxan *Leptomyxa reticulata*; (b) *Zoopage toechospora* within the shelled rhizopod *Euglypha laevis*; (c) *Dactylella tylopaga* within an amoeba. Redrawn from Drechsler, 1935c, 1947a, 1947b; by permission of *Mycologia*

producing a branch within the zone of contact which penetrates the attached host. The animal is prevented from detaching itself by the formation of a swelling on this branch just within the testa. The host is then killed and its protoplasm is assimilated.

Fungi that attack protozoa are widespread in nature and, if the appropriate techniques are used, they can be easily observed, yet nothing is known of their nutritional requirements, physiology, or ecological relationships with their hosts or other micro-organisms. This lack of information is due almost entirely to the fact that none of them has been grown in axenic culture nor in dual culture with suitable hosts. Although some species sporulate abundantly, and their conidia can easily be removed aseptically to nutrient media, the spores do not germinate. Why they do not is unknown, but it has been suggested that this indicates that the fungi are wholly dependent on their hosts for growth and survival. Until more direct evidence is obtained this view cannot be entirely accepted.

2. Rotifer-Trapping Fungi

Three aquatic fungi, *Sommerstorffia spinosa*, *Zoophagus insidians* and *Z. tentaclum*, have been described which capture and consume rotifers (Arnaudow, 1923, 1925; Karling, 1936, 1952; Prowse, 1954). All are Oomycetes, *S. spinosa* belonging to the Saprolegniaceae, while the *Zoophagus* species are probably members of the Pythiaceae, and all three show similar modifications for first capturing rotifers and then invading their bodies.

S. spinosa has a thallus of limited growth which is composed of 3—6 hyphal branches (Figure 4a). Each branch ends in an abruptly tapered adhesive peg, or blunt spine, the contents of which are highly refractive, and it is upon these pegs that rotifers are caught. In the two *Zoophagus* species, hyphae are of indefinite length and may be epiphytic on algae or aquatic fungi. At intervals along the hyphae short, adhesive branches of limited growth arise. In *Z. tentaclum* each branch bears at its apex 1—5 long, fine appendages which are swollen at their bases and are terminated by a small knob (Figure 5a).

All three fungi are host-specific in that they can only capture loricate rotifers, soft-bodied rotifers not normally being caught. When loricate individuals browse along a hypha of *S. spinosa* or *Z. insidians* they are captured on the lateral branches by adhesion of these to their ventral surface. The animal struggles, frequently attempting to bite off the branch, but a crozier of adhesive material is forced between its trophi and firmly attaches the host to the branch. Soft-bodied rotifers can apparently creep over the branches without adhering to them. In *Z.*

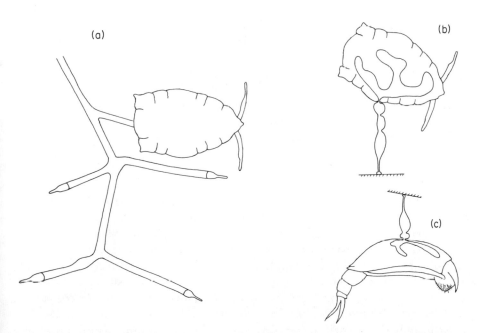

Figure 4 *Sommerstorffia spinosa*: (a) part of thallus showing branches terminating in tapered, adhesive pegs and one captured rotifer: (b) and (c) rotifers captured on sporelings, showing penetration and production of branched, assimilative hyphae. Redrawn from Karling, 1952; by permission of *Mycologia*

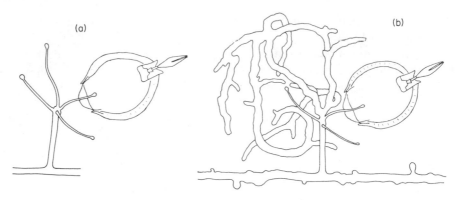

Figure 5 *Zoophagus tentaclum*: (a) lateral branch bearing four filaments terminating in globular, adhesive processes and one captured rotifer; (b) subsequent development of the lateral hyphal branch after capture of a rotifer. Redrawn from Karling, 1936; by permission of *Mycologia*

tentaclum a browsing rotifer becomes captured if it draws one of the knobs that terminate an appendage into its mouth. Adhesion occurs as the animal tries to move away, even though it makes repeated attempts to regurgitate the knob. With all three fungi, if a rotifer is caught it usually dies within 2 hours, possibly as a result of the action of a toxin. In *S. spinosa* and *Z. insidians* the dead animal is then penetrated and a branched, assimilative organ is produced within it. In *Z. tentaclum* the appendage acts as the assimilative organ, remaining little changed as the host's body contents are absorbed. Branches may arise from the appendage-bearing lateral hypha and these grow to enclose the body, but do not penetrate it (Figure 5b).

The behaviour of *S. spinosa* subsequent to the formation of assimilative organs has a number of curious features. Its lobed, assimilative hyphae contain dense cytoplasm and eventually come to fill the whole of the rotifer's body. Their cytoplasm then becomes differentiated into non-flagellate spores which pass to the exterior through wide exit tubes. Each spore gives rise to a biflagellate zoospore which swims, encysts and develops into a small, flask-shaped sporeling anchored to the substratum. The sporeling can trap rotifers by means of its adhesive apex (Figure 4b,c). If no rotifers are captured during the first 8—12 days of sporeling growth, then the latter degenerates and dies. Remarkably, sporelings can also catch the protozoon *Entosiphon ovatum*, which seems to act as an adequate substitute for rotifers.

Although these rotifer-trapping species are by no means uncommon, little is known of their nutrition or physiology. *Z. insidians* can be grown in axenic culture on undefined media, but in nature it, and the other two species, may depend entirely on the exploitation of rotifers for growth and survival. This is implied by the observation that, when a mature thallus of *S. spinosa* captures a host its branches begin to grow, so that it subsequently becomes considerably more extensive. In addition, the degeneration of those sporelings which do not capture a rotifer within a critical period suggests a high degree of obligation to the rotifer-trapping habit. It is worthy of note that at the sporeling stage, which is obviously a critical phase in the life cycle of the fungus, the chances of survival are increased through a widening of normal host range to include a protozoon.

3. Nematode-Trapping Fungi

General Characteristics

There are two basic methods by which active nematode hosts are captured; through adhesion to modified hyphae or by entanglement in non-adhesive hyphal structures. The great majority of nematode-trapping fungi have adhesive snares.

In the Zoopagales there are usually no morphological modifications of the hyphae and all parts of the mycelium are capable of capturing hosts. When a nematode comes into contact with a hypha there is immediately rapid movement of cytoplasmic particles within the hypha at the point where it is touched. This activity quickly spreads along the hypha, and from either side of the contact point particles move towards it through the cytoplasm from some distance away. Within a few seconds a conspicuous droplet of viscous fluid is exuded at the point of contact and this holds the nematode to the hypha (Figure 6a). Despite vigorous movement the animal is rarely able to pull free and, after struggling for 1—2 hours it becomes immobile. The fungus then penetrates its body by means of a fine, lateral branch which is produced at right angles to the main hypha. On entering the body further branches arise from the apex of the lateral and ramify through the nematode (Figure 6a). It is through these that the host's body contents are absorbed, the assimilation process being completed within 24 hours. The hyphae of these fungi are not obviously adhesive until touched but the movement of cytoplasmic particles indicates that the hypha can detect a thigmotropic stimulus and responds with rapid, localized secretion of adhesive material.

In nematode-capturing Hyphomycetes the ability to produce adhesive is usually restricted to specialized, lateral hyphae which are modified in various ways to form morphologically distinct traps. The adhesive is apparently present before nematodes make contact and is in the form of an inconspicuous film over the hyphal surface. Ultrastructural studies show that the cells of adhesive traps contain dense inclusions, not found in the other cells of the mycelium, and abundant rough- and smooth-surfaced membranes (Figures 7 and 8). All these

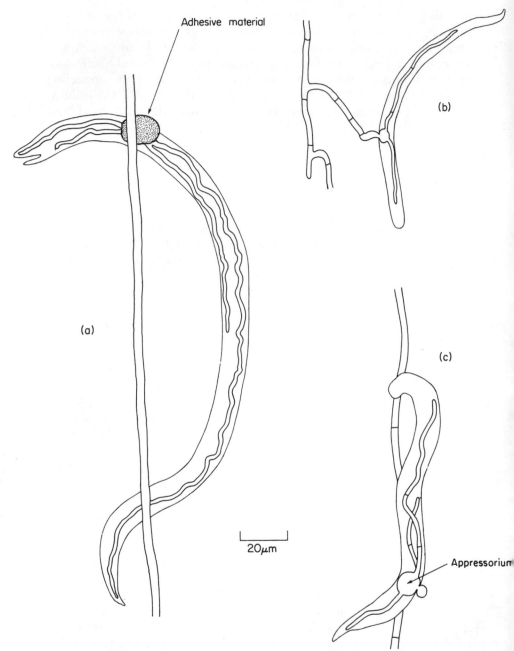

Figure 6   Nematode-trapping fungi: (a) *Stylopage hadra*, dead nematode on adhesive hypha showing assimilative hyphae and accumulation of adhesive material at point of capture; (b) *Triposporina aphanopaga*, nematode captured on undifferentiated, lateral hypha; (c) *Monacrosporium ellipsosporum*, nematode captured on adhesive, lateral branch, showing penetration and production of a globose appressorium within the host. (b) and (c) Redrawn from Drechsler, 1937c; by permission of *Mycologia*

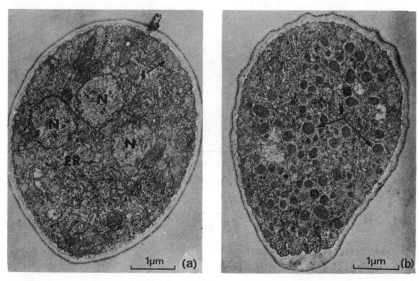

Figure 7   Adhesive   knobs   of   *Monacrosporium   drechsleri*.   Electron-micrographs of sections through adhesive cells showing nuclei (N), dense inclusions (I) and rough endoplasmic reticulum. From Heintz and Pramer, 1972; by permission of the American Society for Microbiology

features are possibly associated with the secretion of adhesive but there are no evident channels in the trap cell walls through which adhesive material might pass (Figure 7a, b).

Each species has only one kind of trap, the most simple, which resemble the main hyphae from which they arise, produce little or no adhesive and are capable of capturing only very small or enfeebled nematodes (Figure 6b). Traps of a greater degree of complexity are derived from two basic structures, these being either lateral branches of very limited growth or branches of potentially unlimited growth. Trap types with limited growth may comprise single-celled laterals or single, globose cells borne upon a pedicel (Figure 9a). Traps with unlimited growth are composed of lateral hyphae which anastomose with both themselves and main hyphae to form two- or three-dimensional networks (Figure 9b,c). All traps are orientated so that they stand up from the surface of the substratum, thus increasing the chance of successful nematode-fungus contact. The pattern of immobilization and exploitation of the host is very similar to that already described for the Zoopagales, except that the assimilative hyphae arise from a spherical appressorium formed within the body of the host (Figure 6c).

Two kinds of non-adhesive trap are known; the passive ring and the constricting ring. The passive ring is a modified hypha of limited growth which, when fully formed, is made up of a slender pedicel bearing a

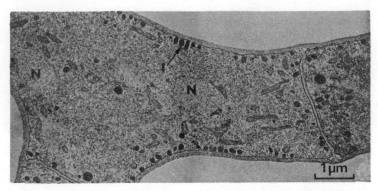

Figure 8   Adhesive network of *Monacrosporium rutgeriensis*. Electronmicrograph of a section through part of an adhesive cell showing nuclei (N) and dense inclusions (I). From Heintz and Pramer, 1972; by permission of the American Society for Microbiology

three- or four-celled annulus (Figure 10a,b). The rings are orientated so that it is possible for a nematode to push its head into one. If the nematode then continues to move forward it will, if it is of a suitable size, become firmly wedged within the ring at some point along its body length. It then struggles, becomes immobile and is invaded from one of the ring cells. Frequently, rings are torn from their pedicels but they are still capable of penetrating the escaped nematode. Nematodes bearing several rings, each representing a successive capture and escape, have been observed, so that death of a contacted host may not be as rapid as with other forms of trap (Figure 10c).

The constricting ring is one of the most remarkable fungal structures known but its physiological mechanisms are poorly understood. It consists of an annulus made up of three arcuate cells supported on a short, stout, one- or two-celled stalk (Figure 10e). When a nematode passes into the ring and touches one of the cells, after a lag of 2–3 seconds, that cell swells rapidly to approximately three times its former volume followed quickly by a similar swelling of the two other cells. The opening of the ring is occluded and the nematode is gripped tightly (Figure 10d). The whole process is irreversible and takes place within the space of 0.1 second. Expansion of the ring cells is accompanied by the appearance of several prominent vacuoles within the centre of each cell, and there is active movement of cytoplasmic particles which does not cease until some time after closure has been completed (Figure 11a–c). After a brief struggle the host dies and is invaded from one of the ring cells. Although ring closure is triggered by a contact stimulus, only the inner surface of the ring cells is sensitive and the increase in cell volume is due to bulging of this inner surface, the outer surface

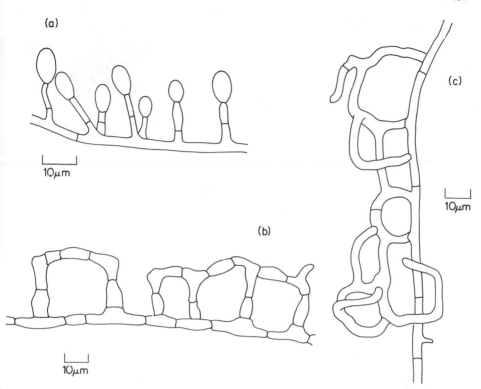

Figure 9   Morphological modifications in lateral adhesive hyphae: (a) *Monacrosporium mutabilis*, terminal globose adhesive cells; (b) *Monacrosporium gephyropagum*, two-dimensional networks; (c) *Arthrobotrys oligospora*, three-dimensional networks. (c) Redrawn from Zopf, 1888

remaining rigid. How the stimulus is perceived is unknown, as is the mechanism by which the ring cells can expand in size so rapidly, but the systems must be very delicate ones as ring cells can be inactivated by periods of ultraviolet irradiation as brief as 60 seconds.

Any theory presented to explain ring closure must take account of a threefold increase in cell volume, involving a 50% increase in cell wall area, taking place in 0.1 second. This obviously requires a rapid change in cell wall structure coupled with a means of inflating the cell. Few critical observations on ring closure have been made and even fewer experimental studies (Couch, 1937; Comandon and de Fonbrune, 1938, 1939; Muller, 1958; Higgins and Pramer, 1967). It is, therefore, not surprising that no single explanation has been put forward which can be accepted without strong reservations. It has been suggested that ring cells contain gelatinous or colloidal compounds, and that closure is brought about by sudden swelling of these, due to either a rearrangement of water molecules within them or to rapid movement of water

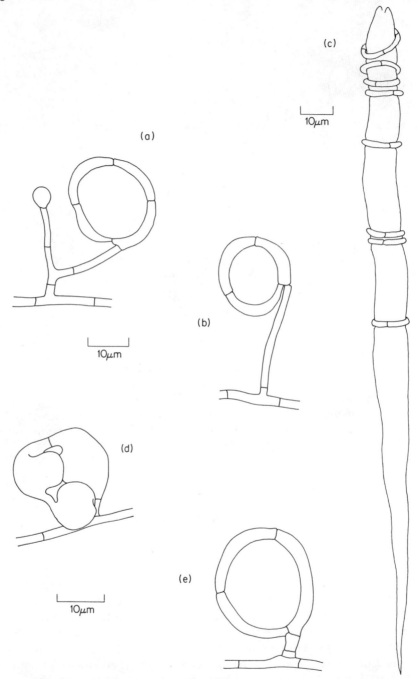

(a)

(b)

(c)

10μm

10μm

10μm

(d)

(e)

Figure 10   Passive and constricting rings: (a) and (b) passive rings; (c) nematode bearing several passive rings torn from their pedicels; (d) and (e) *Arthrobotrys dactyloides*, closed and open constricting rings. (a)–(c) Redrawn from Drechsler, 1937c; by permission of *Mycologia*

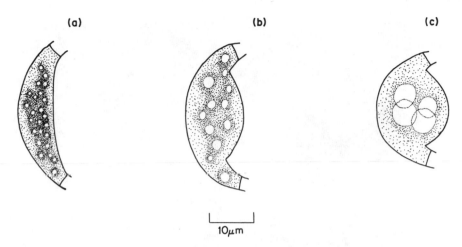

Figure 11   Closure of a constricting ring: (a)—(c) sequence of events in a single ring cells after being triggered, showing bulging of inner walls and development of vacuoles within the cytoplasm. Redrawn from Muller, 1958; by permission of *Transactions of the British Mycological Society*

into them through the stalk cells. There is no direct evidence that either occurs, and with respect to water movement it has been calculated that an average of 18,000 μm^3 of water would have to pass through one or two stalk cells, with cross-sectional areas of only 25 μm^2, in 0.1 second. This is clearly impossible.

It has also been proposed that stimulation causes some form of physical change in the cell wall that leads to a sudden decrease in wall pressure concomitant with an increase in the permeability of the cell membranes to water. The amount of osmotically active substances in the cells then increases, and continues to do so during closure so that water rapidly enters the cell, causing elastic extension of its walls. The inrush of water must presumably be from the surrounding environment rather than from the main hypha via the stalk cells. It is interesting to note that the osmotic pressure of expanded and unexpanded cells, as determined by plasmolysis, is approximately the same. This explanation must be treated with caution, since it is doubtful whether osmotic mechanisms can bring about such very rapid changes in cell volume. It also cannot account for the apparently normal closure of rings which have been formed on aerial hyphae, where there are no obvious water films and where conditions preclude massive movement of water into the ring cells. In addition; no light is shed as to how the closure stimulus is perceived by the sensitive region of the cell wall, nor as to how this stimulus is transmitted to the other cells of the ring if only one of them is stimulated.

Ultrastructural studies on ring cells have demonstrated unique

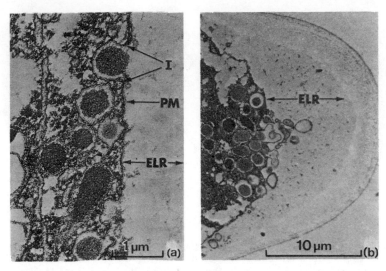

Figure 12   Active rings of *Arthrobotrys dactyloides*. Electron-micrographs of sections through unexpanded ring cells: (a) section towards inner part of cell showing inclusions (I) concentrated immediately beneath the plasma membrane (PM) and the electron —lucent zone (ELR) separating the plasma membrane from the cell wall; (b) entire view of section through luminal border of cell. From Heintz and Pramer, 1972; by permission of the American Society for Microbiology

internal features but have not provided further evidence concerning either mechanisms of stimulus perception or the driving force for cell expansion (Heintz and Pramer, 1972). The unexpanded ring cell protoplast is bounded by a convoluted plasma membrane which forms a labyrinthine matrix outside the cytoplasm (Figure 12). Outside this matrix a wide electron-lucent region abuts on the fibrillar ring cell wall. These features are found only on that side of the ring cell which borders the lumen of the ring. On expansion, the labyrinthine matrix stretches and becomes smooth, the electron-lucent layer disappears and the formerly fibrillar wall becomes non-fibrillar (Figure 13, 14). All these processes allow accommodation of the expanding protoplast.

All nematode-trapping species exploit captured nematodes in the same way, that is by means of a narrow penetration peg giving rise to broader, assimilative hyphae. The fineness of the penetration peg, or if it is relatively wide its snugness of fit in the cuticle, allows the fungus to invade with a minimum of risk that other micro-organisms will also pass into the host's body. Whether penetration is effected by enzymic or physical means is not known. Their mode of nutrition is entirely necrotrophic, extracellular enzymes being produced by the assimilative hyphae within the dead host.

With the possible exception of nematodes caught in passive rings, where immobilization seems to take a relatively long time, host movement usually quickly ceases, with death apparently occurring within 1–2 hours of capture. Capture in a constricting ring obviously badly impairs host movement and function, so that a quick death is to be expected, yet adhesion to a small adhesive cell results in an equally rapid death. It is possible that hosts are killed as a result of mechanical damage due to cuticle penetration, or to the production of the internal appressorium from which assimilative hyphae arise. However, appressoria are not usually formed until the captured animal is moribund or dead, and their diameter is frequently much smaller than that of the nematode (Shepherd, 1955). Furthermore, if a nematode is caught by the tail, the appressorium occurs in a location where physical damage resulting from its formation is minimal, yet death still occurs rapidly. It is likely that these fungi produce toxic substances of some kind which first immobilize and then kill the host. There is limited evidence that, for at least some species, this might be the case. For instance it has been claimed, but not substantiated, that network-forming species secrete nematocidal compounds when grown in liquid media (Soprunov and Galiulina, 1951; Olthof and Estey, 1963). Molecular ammonia has

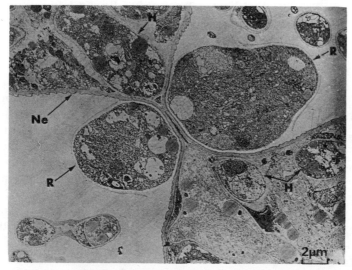

Figure 13 Active rings of *Arthrobotrys dactyloides*. Electronmicrograph of a section through part of a ring and a trapped nematode, showing the severely constricted nematode (Ne) in longitudinal section and two ring cells (R) in cross-section. Sections through hyphae (H) are also present. From Heintz and Pramer, 1972; by permission of the American Society for Microbiology

30

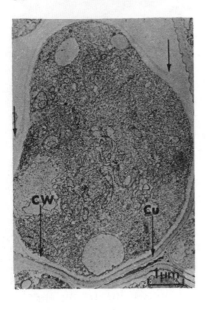

Figure 14 Active rings of *Arthrobotrys dactyloides*. Electronmicrograph of a section through an expanded ring cell showing close contact between the fungal cell wall (CW) and the cuticle of the nematode (Cu). The electron-lucent layer has disappeared and the fungal cell wall has ruptured (arrows). From Heintz and Pramer, 1972; by permission of the American Society for Microbiology

also been proposed as the killing agent in the ring-former *Arthrobotrys dactyloides* (Balan and Gerber, 1972). This species produces ammonia in liquid culture, but then so do many other fungi of all kinds, and it is inconceivable that a single trap and its parent hypha could produce ammonia in sufficient quantity to kill a large and vigorous nematode, particularly in view of the rapid diffusion of ammonia away from the site of synthesis.

If toxins are produced by nematode-trapping fungi, then they are presumably formed by the cells of the traps. Their zone of action must be very localized since nematodes moving across mycelia bearing traps are not immobilized until they are actually caught. Such toxins could be either in the adhesive material or be introduced into the nematode by means of the infection hypha. Much more experimental work is required before the view that toxins are involved in host death can be firmly accepted. It is not easy to envisage what selective advantages there might be for a nematode-trapping fungus to produce an immobilizing toxin. The mechanical strength of even some of the most delicate traps is normally such that a vigorous host has little chance of pulling free and, even when this does happen, then all or part of the trap remains attached to the nematode so that penetration from this is usually possible.

Most species of nematode-trapping Hyphomycetes, but not the Zoopagales, are easily brought into axenic culture where, in the absence of nematodes, they commonly do not form traps. On addition of nematodes to cultures traps are formed within 24 hours. This behaviour is not entirely unexpected, since trap formation requires expenditure of energy and to do this in the absence of potential hosts would be an uneconomic activity. An obvious selective advantage would accrue to a fungus if it only produced traps as a response to the presence of nematodes. In axenic culture it has been demonstrated that water in which nematodes have been incubated, or cell-free nematode extracts, induce trap formation, as does animal material other than nematodes. Using the network-forming species *Arthrobotrys conoides* as a test fungus it has been shown that broths in which the animal-parasitic nematode *Neoaplectana glaseri* has developed axenically contain a water-soluble morphogenic substance, termed 'nemin', which can induce trap formation within 24—48 hours of application (Pramer and Stoll, 1959).

It is clear that nemin is not a single compound and it is probable that a number of compounds, perhaps closely related biochemically, are responsible for morphogenesis of traps. It has been proposed that proteins normally excreted by nematodes are ultimately responsible (Jackson, 1961; Pramer and Kuyama, 1963). These might then be hydrolysed at the fungal cell surface and the relatively low molecular weight peptides so released may be the actual morphogenetic compounds (Nordbring-Hertz and Brink, 1974). In support of this view it has been shown that active compounds can be obtained by enzymatic hydrolysis of ascarid proteins and that these are capable of passing through dialysis film. However, none of the large number of proteins, peptides, and amino acids so far tested will induce trap formation in *A. conoides*. This contrasts with reports that valine, leucine and isoleucine stimulate trap formation in the same fungus. (Wooton and Pramer, 1966). Carbohydrates such as arabinose, ribose, and rhamnose, as well as glycerol, have been reported as stimulating *A. conoides* to produce traps (James and Nowakowski, 1968). It has also been shown that nemin-induced trap formation in *A. conoides* is CO_2-dependent, the fungus producing no traps in its absence (Bartnicki-Garcia, Eren and Pramer, 1964).

Although trap formation as a response to exogenous biochemical stimuli undoubtedly takes place, and probably in some circumstances has great biological and ecological significance, species differ in the degree of their response to nematode extracts (Feder, Everard and Wooton, 1963). In addition, different isolates of a single species frequently show a range of behaviour with respect to their readiness to form traps (Lawton, 1957; Cooke, 1963c, 1964; Cooke and Satchuthananthavale, 1966). At the extremes, isolates may either not form traps at all, even in the presence of live nematodes, or form traps spontaneously and abundantly in the absence of nematodes. It is quite clear then that morphogenesis of traps is not simply a response to exogenous stimuli.

Ecological Characteristics, Nutrition and Physiology

The nematode-trapping fungi have been known for a considerable period of time but the basis for the existence of this habit is not understood. An obvious area in which to seek information that might clarify the problem is that of the ecology of these fungi. They are present and active in decaying organic matter and can easily be isolated from soil, where they presumably occupy microhabitats consisting of organic matter particles. The great majority of them are not obligately nematode-trapping since they can be grown in axenic culture, the exceptions being species of Zoopagales. It nevertheless seems likely that lack of success with the latter fungi is due to technical reasons rather than to their being absolutely dependent on nematodes.

Three basic approaches have been made in studies on the ecology of these fungi. First, attempts have been made to gain an insight into the significance of the trapping habit by studying cultures of them in association with populations of suitable host nematodes. Second, attempts have been made to directly observe their behaviour in soil under experimentally varied conditions. Finally, axenic cultures of a number of species have been studied, with emphasis on defining those characteristics that might have ecological importance when the fungi are in their natural habitats.

Studies involving cultures of the fungi in association with suitable host populations have proved to be unrewarding. When colonies growing on agar plates are presented with adequate numbers of nematodes the fungi may respond by quickly producing traps, but the duration of trapping activity is usually relatively short (Feder, 1963). A similar sequence of events occurs if nematode-trapping fungi are allowed to grow out from natural soil inocula containing nematodes. This restriction of activity in time suggests that the nematode-trapping habit supplements the nutritional needs of these fungi rather than supplying all the nutrients that they require. This view is supported by a number of observations (Hayes and Blackburn, 1966; Olthof and Estey, 1966; Monoson, 1968). Comparisons of extension rates of colonies growing in the presence and absence of nematodes show that the exploitation of nematodes has no consistent effect on the growth rate of any fungus so far studied. Experiments in which fungi have been added to sterile sand supplemented with nutrient solutions of various kinds indicate that nematodes alone cannot provide all the energy required for growth.

Further evidence supporting the view that these fungi cannot grow or survive solely through trapping nematodes has been obtained from observing their behaviour in non-sterile soil (Cooke, 1962a,b, 1963a). In soil, nematode-trapping fungi are closely associated with decomposing organic substrates as are soil nematodes which are either feeding

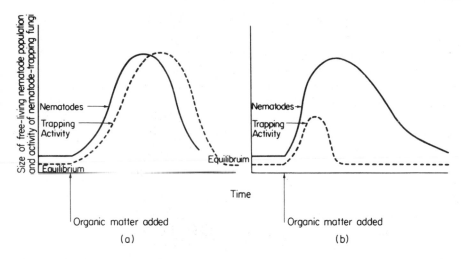

Figure 15　Relationship between a free-living nematode population in the soil and nematode-trapping fungi after addition of organic matter to the soil: (a) expected changes in nematode population and trapping activity if a labile equilibrium exists between the two groups of organisms; (b) actual pattern of change

on organic matter directly or upon other micro-organisms that are in some way themselves dependent on it. Soil nematode populations may therefore be increased by adding living or dead plant tissues to the soil. If the fungi were completely dependent on capturing nematodes for growth and survival then such amendment of soil should cause their numbers or activity to increase as the nematode population increases (Figure 15a). Their numbers or activity should continue to increase until the nematodes, being destroyed more rapidly than they can multiply, begin to decline in number. With this decline the activity of the fungi should also then decline, until an equilibrium point is reached corresponding more or less to that existing before organic amendment took place. Observed changes in trapping activity and nematode population subsequent to organic matter amendment of soil do not, however, fit this pattern (Figure 15b). In unamended soil the nematode population remains low and trapping activity is normally at an undetectable level. In soil amended with green plant tissue the nematode population follows the expected rise and fall pattern and so does trapping activity. However, the two are in no way simply related. Trapping activity is frequently declining while the nematode population is either rising or is still at a high level. Since a simple labile equilibrium between the fungi and nematodes does not exist, the nematode-trapping habit seems to be in some way involved in the supplementation of saprotrophic activity.

An examination of nematode-trapping species in axenic culture shows that they frequently differ from one another in a number of basic characteristics, and this probably means that they differ too in their behaviour in nature. Four of these characteristics which have been studied are rate of mycelial extension, ability to produce traps spontaneously in the absence of nematodes, competitive saprotrophic ability, and nematode-trapping efficiency (Cooke, 1963a,b, 1964). In general, although there are some exceptions, species that have adhesive networks have the most rapid growth rates. In contrast, species with adhesive branches or non-constricting or constricting rings have much slower growth rates. Slow growth rate seems to be associated with the ability to produce traps spontaneously, network-formers doing so rarely while other species do so frequently, particularly on germlings arising from germinating conidia. There is also a relationship between trap type and competitive saprotrophic ability. This characteristic has been studied only in network-forming and constricting ring-forming species, the former being good competitors with the natural soil flora while the latter are not. When cultures of nematode-trapping fungi are added to non-sterile soil the effect that they have on the indigenous nematode population depends on what kind of traps the fungi produce. Ring- and adhesive branch-forming species seem to be much more efficient in reducing soil nematode populations, during the short period of trapping which occurs after their addition, than are network-formers.

If the trapping habit represents a series of adaptations allowing an easing of, or escape from, competition for available substrates during decomposition of organic matter, then it might follow that species which are more susceptible to the rigours of competition, in contrast to species better equipped to compete, would be highly efficient in trapping nematodes and have a well-developed ability to produce traps even when stimuli from nematodes were weak. Using these criteria, together with the observed differences in growth rate and competitive saprotrophic ability, then the ring-formers and some of the adhesive branch-formers may be more advanced towards an obligately nematode-trapping mode of life than other nematode-trapping fungi. This is, of course, conjecture, but evidence from nutritional studies indicates that it may be correct.

In axenic culture nematode-trapping fungi do not exhibit a unique nutritional pattern but closely resemble many saprotrophic fungi in having relatively simple requirements (Coscarelli and Pramer, 1962; Grant, Coscarelli and Pramer, 1962; Faust and Pramer, 1964; Blackburn and Hayes, 1966; Satchuthananthavale and Cooke, 1967b). They are able to utilize a wide range of simple carbohydrates as sole sources of carbon and a wide range of nitrogen compounds as sole sources of nitrogen. However, differences occur within the nematode-trapping Hyphomycetes with respect to the ability of species with different trap types to utilize various carbon and nitrogen compounds. Detailed studies have been made on species having one of two trap types, either adhesive networks or constricting rings (Satchuthananthavale and Cooke, 1967a,b,c). In comparison with the network-formers, ring-forming fungi show reduced nutritional versatility. They do not readily utilize carbohydrates more complex than hexoses and lack the ability to utilize nitrite and nitrate nitrogen, although network-forming fungi will do so.

This loss of nutritional versatility could be associated with a high degree of dependence upon the nematode-trapping habit for survival; for example, ring-forming fungi and other species with diminished versatility may obtain the bulk of their essential nitrogenous materials from their nematode hosts, while species that readily utilize inorganic nitrogen sources may not be so dependent. An inability to utilize carbon compounds other than relatively simple, and in the soil ephemeral, ones may mean that natural activity of the ring-formers is severely restricted in space and time. Activity would only be possible where and when such readily utilizable energy sources were available, although it is possible that nematodes provide them with sufficient carbon compounds to allow them to be active for longer periods. It is clear from ecological and nutritional studies that not all nematode-trapping fungi are capturing nematodes for the same reason.

Biological Control

Many attempts have been made to reduce populations of soil-borne plant-parasitic nematodes by using nematode-trapping fungi. The results and implications of these attempts have been reviewed in detail a number of times (Duddington, 1957, 1962; Cooke, 1968; Duddington and Wyborn, 1972). Organic amendments of different kinds, cultures of the fungi, or both, have been added to soil prior to the planting of a susceptible crop. There has been no consistent success and the great majority of experiments have failed to achieve even a modest degree of control. The basis for attempted control has always been the assumption that a labile equilibrium exists between the fungi and nematodes. Since this is obviously not so it is apparent why, in the main, biological control measures have proved useless. Any period of activity which does result from soil amendment does not last long enough, or is not intense enough, to reduce nematode populations significantly. In addition, when nematode-trapping species are introduced into soil they are subject to intense antagonisms from other soil micro-organisms and there is usually pronounced lysis of hyphae and conidia within a very short time (Tarjan, 1961; Mankau, 1962; Cooke and Satchuthananthavale, 1968).

Extravagant claims have been made for the effectiveness of these

36

fungi in controlling nematode pests of animals. In particular, it has been claimed that ancylostomiasis in coal miners can be reduced by treating infested mine workings with conidia (Soprunov, 1966). Such statements should be treated with reserve.

Bibliography

Arnaudow, N., 1923. *Flora, Jena,* 116: 109–113.
Arnaudow, N., 1925. *Flora, Jena,* 118–9: 1–16.
Balan, J. and Gerber, N. N., 1972. *Nematologica,* 18: 163–175.
Bartnicki-Garcia, S., Eren, J. and Pramer, D., 1964. *Nature, Lond.* 204: 804.
Blackburn, F. and Hayes, W. A., 1966. *Ann. appl. Biol.,* 58: 43–50.
Comandon, J. and de Fonbrune, P., 1938. *C. r. Soc. Biol.,* Paris. 129: 620–625.
Cooke, R. C., 1962a. *Trans. Br. mycol. Soc.,* 45: 314–320.
Cooke, R. C., 1962b. *Ann. appl. Biol.,* 50: 507–513.
Cooke, R. C., 1963a. *Nature, Lond.,* 197: 205.
Cooke, R. C., 1963b. *Ann. appl. Biol.,* 51: 295–299.
Cooke, R. C., 1963c. *Ann. appl. Biol.,* 52: 431–437.
Cooke, R. C., 1964. *Ann. appl. Biol.,* 54: 375–379.
Cooke, R. C., 1968. *Phytopathology,* 58: 909–913.
Cooke, R. C. and Satchuthananthavale, V., 1966. *Trans. Br. mycol. Soc.,* 49: 27–33.
Cooke, R. C. and Satchuthananthavale, V., 1968. *Trans. Br. mycol. Soc.,* 51: 555–561.
Coscarelli, W. and Pramer, D., 1962. *J. Bact.,* 84: 60–64.
Couch, J. N., 1937. *J. Elisha Mitchell scient. Soc.,* 52: 301–309.
Drechsler, C., 1934. *J. Wash. Acad. Sci.,* 24: 395–402.
Drechsler, C., 1935a *Mycologia,* 27: 6–40.
Drechsler, C., 1935b. *Mycologia,* 27: 176–205.
Drechsler, C., 1935c. *Mycologia,* 27: 216–223.
Drechsler, C., 1936. *J. Wash. Acad. Sci.,* 26: 397–404.
Drechsler, C., 1937a. *Mycologia,* 29: 229–249.
Drechsler, C., 1937b. *J. Wash. Acad. Sci.,* 27: 391–398.
Drechsler, C., 1937c. *Mycologia,* 29: 447–552.
Drechsler, C., 1941. *Biol. Rev.,* 16: 265–290.
Drechsler, C., 1946. *Mycologia,* 38: 1–23.
Drechsler, C., 1947a. *Mycologia,* 39: 253–281.
Drechsler, C., 1947b. *Mycologia,* 39: 379–408.
Duddington, C. L., 1955. *Bot. Rev.,* 21: 377–439.
Duddington, C. L., 1956. *Biol. Rev.,* 31: 152–193.
Duddington, C. L., 1957. *The Friendly Fungi,* Faber and Faber, London.
Duddington, C. L., 1962. *Predaceous Fungi and the Control of Eelworms. Viewpoints in Biology,* Vol. 1, Butterworth, London.
Duddington, C. L. and Wyborn, C. H. E., 1972. *Bot. Rev.,* 38: 545–565.
Faust, M. A. and Pramer, D., 1964. *Life Sci.,* 3: 141–143.
Feder, W. A., 1963. *Mycopath. Mycol. appl.,* 19: 99–104.
Feder, W. A., Everard, C. O. R. and Wooton, L. M. O., 1963. *Nematologica,* 2: 49–54.
Grant, C. L., Coscarelli, W. and Pramer, D., 1962. *Appl. Microbiol.,* 10: 413–417.
Hayes, W. A. and Blackburn, F., 1966. *Ann. appl. Biol.,* 58: 51–60.
Heintz, C. E. and Pramer, D., 1972. *J. Bact.,* 110: 1163–1170.
Higgins, M. L. and Pramer, D., 1967. *Science. N.Y.,* 155: 345–346.

Jackson, G. J., 1961. *Expl. Parasit.*, 11: 241—247.

James, A. W. and Nowakowski, R. J., 1968. *Can. J. Microbiol.*, 14: 1260—1261.

Karling, J. S., 1936. *Mycologia*, 28: 307—320.

Karling, J. S., 1952. *Mycologia*, 44: 387—412.

Lawton, J. R., 1957. *Trans. Br. mycol. Soc.*, 50: 195—205.

Mankau, R., 1962. *Phytopathology*, 52: 611—615.

Monoson, H. L., 1968. *Mycologia*, 60: 788—801.

Muller, H. G., 1958. *Trans. Br. mycol. Soc.*, 41: 341—364.

Nordbring-Hertz, B. and Brink, C., 1974. *Physiologia Pl.*, 31: 59—63.

Olthof, Th. H. A. and Estey, R. H., 1963. *Nature, Lond.*, 197: 514—515.

Olthof, Th. H. A. and Estey, R. H., 1966. *Nature, Lond.*, 209: 1158.

Pramer, D. and Kuyama, F., 1963. *Bact. Rev.*, 27: 282—292.

Pramer, D. and Stoll, N. R., 1959. *Science, N.Y.*, 129: 966—967.

Prowse, G. A., 1954. *Trans. Br. mycol. Soc.*, 37: 134—150.

Satchuthananthavale, V. and Cooke, R. C., 1967a. *Nature, Lond.*, 214: 321—322.

Satchuthananthavale, V. and Cooke, R. C., 1967b. *Trans. Br. mycol. Soc.*, 50: 221—228.

Satchuthananthavale, V. and Cooke, R. C., 1967c. *Trans. Br. mycol. Soc.*, 50: 423—428.

Shepherd, A. M., 1955. *Nature, Lond.*, 175: 475.

Soprunov, F. F., 1966. *Predaceous Hyphomycetes and their Application to the Control of Pathogenic Nematodes*, Translation Publ. for the U.S.D.A. and National Science Foundation, Washington.

Soprunov, F. F. and Galiulina, E. A., 1951. *Mikrobiologiya*, 20: 489—499.

Tarjan, A. C., 1961. *Soil Crop. Sci. Soc. Florida Proc.*, 21: 17—36.

Wooton, L. M. O. and Pramer, D., 1966. *Bact. Proc.*, p. 31.

Zopf, W., 1888. In A. Schenk, *Handbüch der Botanik*, Breslau.

Chapter 3

Endosymbiotic Fungi of Microscopic Animals

A great many species of fungi from widely separated taxonomic groups spend the major part of their life cycle within the bodies of microscopic animals. When infected by a biotrophic endosymbiotic species an animal may remain alive until a relatively late stage in the development of the fungus. In contrast, if an animal becomes infected by a necrotrophic fungus death occurs rapidly, frequently during the very early stages of penetration, and the fungus then develops within the moribund body. There are numerous, widely scattered reports of fungi living within eggs or cysts of microscopic animals, but these will not be discussed here as so little is known of them. Most detailed studies on endosymbiotic fungi have been concerned with species that destroy protozoa, rotifers and nematodes.

1. Endosymbiotic Fungi of Protozoa and Rotifers

Although a large number of fungi of protozoa have been described, details of their biology are not well known. This is due mainly to the relative rarity with which they are encountered and to their minute size. They have not yet been grown in axenic culture, nor in dual culture with suitable host animals, so that no experimental studies of them have been made. The negative results of attempts at axenic culture imply that they are obligately symbiotic but evidence for this is only circumstantial. Despite the difficulties arising from lack of cultural studies much can be inferred of the biology of these organisms from direct microscopic examination of infected hosts.

All species so far described are members of the Zoopagales and many of them are supreme examples of biotrophy, being able to exploit, often over relatively long periods, a minute unicellular host. Infection of the host is effected when it either ingests fungal spores or when these adhere to the body surface.

The majority of known endosymbiotic fungi of protozoa are species of *Cochlonema*, most of which are associated with amoebae, although a few exploit shelled rhizopods. It is not clear whether *Cochlonema*

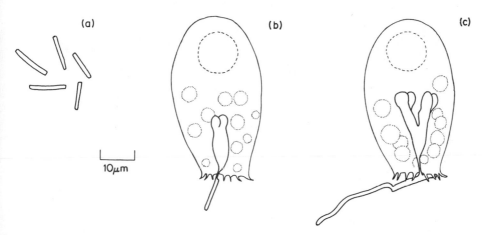

Figure 16   *Cochlonema explicatum*: (a) adhesive conidia; (b) infection of the shelled rhizopod *Sphenoderia dentata*, showing adhesion of the conidium and production of an assimilative branch; (c) production of an external hypha by the infecting conidium. Redrawn from Drechsler, 1955; by permission of *Mycologia*

species, or other endosymbiotic fungi attacking protozoa, are host specific, but it is almost certain that those species invading amoebae do not infect shelled rhizopods. Of the few *Cochlonema* species that have shelled rhizopods as their hosts only one, *C. explicatum*, has adhesive conidia, the remainder gaining entry when their spores are ingested (Drechsler, 1955). The conidium of *C. explicatum* adheres to the protoplasm at the aperture of the shell and, upon germination, gives rise to a penetration hypha which widens within the host and grows to form a large, dichotomously branched thallus. The infecting spore is not shed and is supplied with nutrients by the developing thallus (Figure 16a–c). *C. explicatum* is unique in that, unlike other endosymbiotic fungi of protozoa, it can produce an extensive external mycelium. There is a striking resemblance between this internal thallus and the assimilative organs of fungi that capture protozoa. In *Cochlonema* species whose spores are ingested, there is rapid germination and each conidium gives rise to a curved or coiled thallus (Drechsler, 1959). The thallus grows, the body contents of the host are absorbed and, after its death, hyphae of limited growth emerge through the aperture in the testa and become transformed to chains of conidia.

All *Cochlonema* species of shelled protozoa bring about rapid death of the host and appear to be necrotrophs. In contrast, *Cochlonema* species that infect amoebae are biotrophic and most infect through ingested spores (Drechsler, 1941a, 1946b). The ingested conidium germinates and its germ tube gives rise to a coiled, often dichotomously branched thallus (Figure 17a, b). The thallus increases in size at the expense of host protoplasm but the host remains alive for a relatively

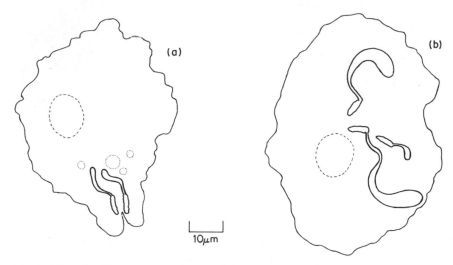

(a)

(b)

10μm

Figure 17   *Cochlonema symplocum*: (a) germination of ingested conidia within an amoeba; (b) differentiation of curved thalli from the germ tubes. Redrawn from Drechsler, 1941a; by permission of *Mycologia*

long period during thallus growth. Multiple infections commonly occur as a result of the ingestion of more than one conidium, the host being able to support several thalli. In *C. agamum* the cytoplasm of the host can be reduced by about 75% before the animal becomes moribund. There is gradual slowing of movement, the contractile vacuole ceases to function, and the nucleus begins to degenerate. If infected hosts are placed at 15°C, which is below the optimum temperature for thallus growth, the latter ceases to develop and becomes vacuolate. The host's contractile vacuole then regains its function and the host nucleus assumes its normal appearance (Drechsler, 1946b). When the host finally becomes sluggish, but before its death, asexual reproduction of the fungus is initiated and hyphae arise from thalli, pass to the exterior and become transformed to chains of conidia. Tolerance of the host to multiple infection may be associated with heterothallism in the fungus so that in order for sexual reproduction to take place sexually compatible thalli must be present in the host. However, some species that normally cause multiple infections are homothallic.

In *Endocochlus*, a biotrophic genus closely related to *Cochlonema*, there is also a fine balance between development of the fungus and the running down of the host cell (Drechsler, 1935). All *Endocochlus* species have adhesive conidia which may become attached to the surface of an amoeba. On germination a wide, median appressorium is formed and from this a fine, conical penetration tube passes into the host's protoplasm (Figure 18a–b). After gaining entry the tip of the penetration tube swells to form a spherical body which becomes filled with cytoplasm that migrates to it from the conidium (Figure 18c). This body breaks off, becomes elongate, coils, and may finally be dichotomously branched (Figure 18d–f). The empty conidium and penetration tube are then shed. Multiple infections are normal and several thalli may be supported by a single host. As with *Cochlonema* infections, the host finally becomes sluggish, and when it begins to degenerate, or sometimes a little before this, asexual reproduction takes place.

Although amoebae containing spores or young thalli of *Cochlonema* and *Endocochlus* can sometimes expel their fungi, the majority of infections are successful. That most thalli, even juvenile ones, remain within host protoplasm and reach maturity indicates that they have special characteristics which enable them to do so. After being ingested by amoebae, spores of other fungi, the remains of algae, and the shells of small rhizopods, are normally subsequently cast out without difficulty. Thalli may, therefore, possess some intrinsic qualities which obviate this. It is interesting to note that, occasionally, adhesive conidia of *Endocochlus* may be ingested but they then germinate normally. The

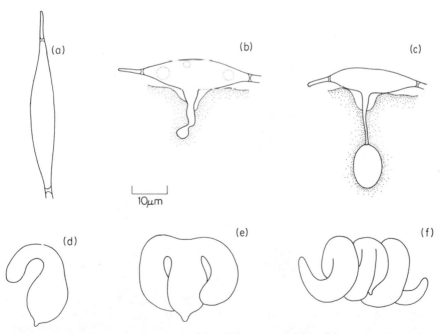

Figure 18 *Endocochlus binarius*: (a) adhesive conidium; (b) penetration of an amoeba by a germ tube arising from a conidium; (c) young thallus forming at tip of penetration tube; (d)–(f) development of coiled thallus. Redrawn from Drechsler, 1949; by permission of *Mycologia*

empty conidium and germ tube are ejected by the animal, but the young thallus remains within the host's cytoplasm. As well as this ability to remain within the cytoplasm there must be a degree of balance between fungus and host. This implies a lack of those fungal enzymes that might otherwise kill the host rapidly, together with some kind of resistance on the part of the fungus to those host enzymes which might normally bring about dissolution of fungal cell walls and cytoplasm.

Two further genera, *Amoebophilus* and *Bdellospora* exploit amoebae but, in contrast to *Cochlonema* and *Endocochlus*, their thalli remain outside the host, which is penetrated by an absorptive organ (Drechsler,

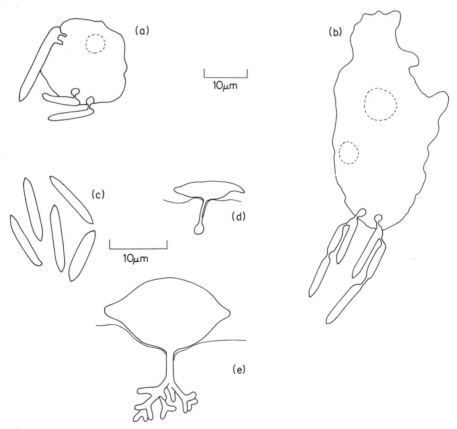

Figure 19   *Amoebophilus sicyosporus* and *Bdellospora helicoides*: (a) germination of *A. sicyosporus* conidia on an amoeba and formation of globular assimilative processes; (b) external thalli of *A. sicyosporus* with conidial chains; (c) adhesive conidia of *B. helicoides*; (d) penetration of an amoeba by a germinating conidium; (e) development of an external thallus and a branched assimilative hypha. Redrawn from Drechsler, 1935, 1959; by permission of *Mycologia*

1935, 1959). Although they are thus not strictly endosymbiotic they are clearly closely related to endozoic fungi, particularly *Endocochlus* species. The absorptive organ can be considered to be equivalent to an internal thallus, while the external thallus is derived directly from an adhesive conidium. The latter germinates by means of a fine penetration tube, the tip of which then grows to form a lobed or branched absorptive organ which remains very small in size in relation to the body of the host (Figure 19c–e). The conidium then swells and becomes transformed into a small thallus. In *Amoebophilus* the conidium, or thallus, never increases in size but instead short chains of spores are produced by means of its apical growth (Figure 19a, b). In *Bdellospora* the thallus does increase in size, but long chains of adhesive conidia are also produced from it. Asexual reproduction in both fungi is initiated while the host is still alive and mobile, but as sporulation continues there is an increased demand for host nutrients, so that host movement gradually slows. Heterothallic sexual reproduction then takes place, perhaps being initiated by the change in nutrient status of the host, and it is during this sexual phase that the host finally dies. The possession of an external thallus minimizes the risk of ejection by the host, and the minuteness of the absorptive organ allows minimum interference with host function until the initiation of asexual reproduction.

A number of endosymbiotic fungi destroy rotifers but there is little information available as to their biology although they seem to be biotrophs and their thalli develop within the host while the latter is still alive. (Karling, 1944, 1946).

2. Nematode-Destroying Fungi

Many endosymbiotic fungi destroy nematodes and, in contrast to those of protozoa, most are necrotrophic (Cooke and Godfrey, 1964). As is the case with the nematode-trapping mode of life the endosymbiotic habit has arisen independently within widely different groups of fungi.

General Characteristics

Endosymbiotic fungi of nematodes are of two kinds. A relatively small number of species, almost all Chydridiomycetes or Oomycetes, have internal thalli that are either unicellular or are made up of hyphal bodies loosely aggregated together. In contrast, the majority of species, normally Hyphomycetes and species of imperfect Basidiomycotina, have a thallus consisting of a small mycelium of assimilative hyphae.

During development non-mycelial thalli increase in size until they occupy almost the whole of the body of the host which at this stage is usually dead. Motile or non-motile spores, which are the agents of infection, are then produced. In some fungi these are formed endogenously within the thalli and are then released, or forcibly ejected, through exit tubes or pores passing to the exterior (Figure 20a). Infection by non-motile spores is usually through their adhesion to the host's cuticle upon which they germinate, entering the body directly by means of a fine penetration tube. Motile spores may cluster around the mouth or excretory pore of the host where they then encyst and, in some way not yet established, bring about infection. Adult nematodes may be more susceptible to infection through zoospore contact than are juveniles (Sayre and Keeley, 1969). Many, but not all, of these non-mycelial fungi seem to be biotrophs and during thallus growth the host remains alive, although it moves increasingly slowly as development takes place. Hosts can tolerate the presence of many thalli derived from either multiple infection or from the division of the original thallus (Figure 20b). Simultaneous multiple infection of a single host by two different fungi can occur (Figure 21). There is no evidence that any of the known species are host-specific.

Species producing an endosymbiotic mycelium are invariably necrotrophic and may be divided into two biological groups according to the manner in which they infect the host. Members of the first group have

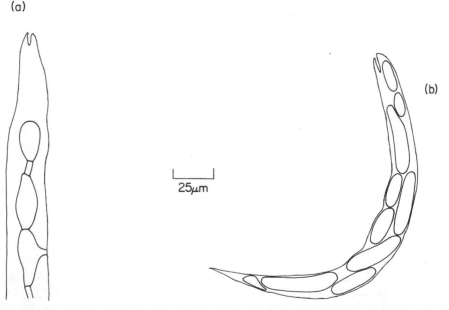

(a)

(b)

25μm

Figure 20   Endosymbiotic fungi of nematodes: (a) thallus of *Catenaria anguillulae* within a dead nematode, one hyphal body with an exit tube; (b) thalli of *Gonimochaete horridula* within a nematode.(a) Redrawn from Sorokine, 1876

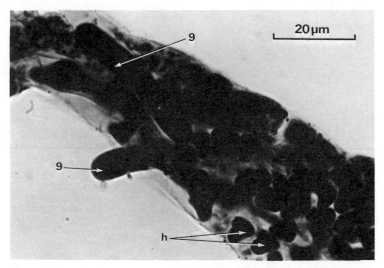

Figure 21   Multiple infection by endosymbiotic fungi. Part of dead body of a nematode containing spores of *Haptoglossa heterospora* (h) and thalli of *Gonimochaete horridula* (g)

adhesive conidia with a wide range of shape depending on the species. They are either entirely coated with an adhesive film or secrete a minute droplet of adhesive fluid from a glandular protuberance formed at one end. Adhering spores germinate on the host's cuticle which is breached by means of a fine penetration tube. This gives rise to an internal assimilative mycelium that rapidly fills the host. After death of the nematode, hyphae of limited growth emerge from the body and upon these are borne further conidia (Figure 22a). In some species of one genus, *Nematoctonus*, these fertile hyphae may also bear short, adhesive cells that can capture passing nematodes (Figure 22b).

The second group of mycelial, endosymbiotic fungi have conidia that are ingested by nematodes (Aschner and Kohn, 1958; Barron, 1970). This group comprises species of a single hyphomycetous genus, *Harposporium*, and their spores may be filiform, pod-shaped, arcuate, anvil-shaped, or barbed (Figure 23a–c). When such spores are ingested these various morphological modifications cause them to lodge in the buccal cavity, the oesophageal bulb, or even the hind end of the gut. Here they germinate and hyphae ramify through the host's body, eventually filling it. After death of the nematode, fertile hyphae emerge from it and sporulation takes place (Figure 24a, b). Adhesive-spored fungi are not host-specific but those species whose conidia must be ingested are specific in that nematodes which have their mouthparts modified to form a stylet cannot be infected. Infected nematodes with these sucking mouthparts have been observed occasionally but the

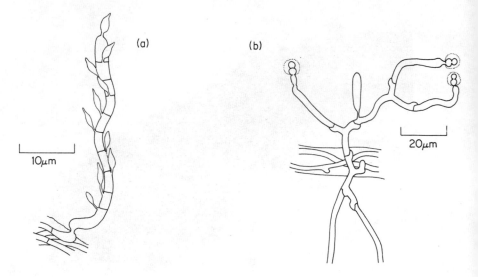

Figure 22   Endosymbiotic Hyphomycetes of nematodes: (a) *Meria coniospora*, conidiophore bearing adhesive conidia arising from a dead nematode; (b) *Nematoctonus haptocladus*, hyphae with clamp connections arising from a dead nematode and bearing one conidium and three adhesive, glandular cells. Redrawn from Drechsler, 1941*b*; by permission of *Phytopathology*; Drechsler, 1946a; by permission of *Mycologia*

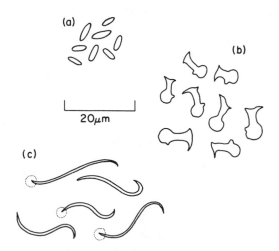

Figure 23   Range of size and shape of *Harposporium* conidia: (a) *H. baculiforme*; (b) *H. dicorymbum*; (c) *H. helicoides*. (b) and (c) redrawn from Drechsler 1941*b*, 1963; (b) by permission of *American Journal of Botany*; (c) by permission of *Phytopathology*

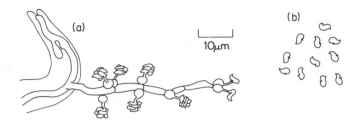

Figure 24 *Harposporium diceraeum*: (a) fertile hyphae with conidia arising from a dead nematode; (b) barbed conidia. Redrawn from Drechsler, 1941*b*; by permission of *Phytopathology*

fungus must have entered fortuitously, perhaps through the mouth or more probably through another body opening (Capstick, Twinn and Waid, 1957; Duddington and Wyborn, 1972).

Death of the host is usually rapid and even in instances where death occurs after a relatively long period this is due to slow development of the fungus rather than to a compatibility between it and the host's living tissues.

Nutrition and Physiology

Diseased nematodes can be separated from soil or other substrata and the development of fungi within them can be observed. Spores can also be separated from soil and used to either infect cultured nematode populations or to establish axenic cultures of the fungi (Barron, 1969; Giuma and Cooke, 1972). Many mycelial species have been brought into pure culture but few with non-hyphal thalli have been grown apart from their hosts, although they can be maintained in dual culture (Davidson and Barron, 1973). The notable exceptions are species of *Catenaria*, but members of this genus are probably mainly saprotrophic on dead or moribund nematodes or other microscopic animals rather than being necrotrophic endosymbiotic fungi. The bulk of present knowledge of the nutrition and physiology of endosymbiotic nematode-destroying fungi is, therefore, derived from investigations made on axenic cultures of mycelial forms, in particular species of *Harposporium* and *Nematoctonus*.

When developing within and sporulating upon their hosts, *Harposporium* and *Nematoctonus* species show limited mycelial growth and this, together with their obviously extreme physical specialization, has been taken to reflect the possession of peculiar or exacting nutrient requirements. Surprisingly, this is not so, and species of *Harposporium* will grow well on a simple glucose—mineral salts medium, though some isolates may have a partial requirement for thiamine. They can utilize a wide range of carbohydrates, including polysaccharides, as sole sources of carbon, and a wide range of nitrogen compounds as sole nitrogen sources. They may, however, grow poorly on non-reduced forms of nitrogen (Aschner and Kohn, 1958; Duddington and Wyborn, 1972). The nutritional requirements of *Nematoctonus* species resemble those of *Harposporium* very closely, except that there is not even a partial requirement for vitamins, but they do require an unidentified non-vitamin factor present in yeast extract (Bricklebank and Cooke, 1969; Giuma and Cooke, 1974a). Species of both genera grow well with glycogen or glycerol as sole carbon source and this may be related to their ability to utilize the glycogen and lipid-rich bodies of nematodes.

This lack of nutritional specialization suggests that, at least in *Harposporium* and *Nematoctonus*, the endosymbiotic habit in mycelial fungi is associated with an inability to compete with other micro-organisms rather than with a biochemical deficiency of any sort. There still exists, however, the possibility that these fungi are completely dependent on nematodes for reduced forms of nitrogen.

Some endosymbiotic species that have adhesive conidia produce nematotoxic compounds (Giuma and Cooke, 1971; Giuma, Hackett and Cooke, 1973). For example, conidia of some *Nematoctonus* species

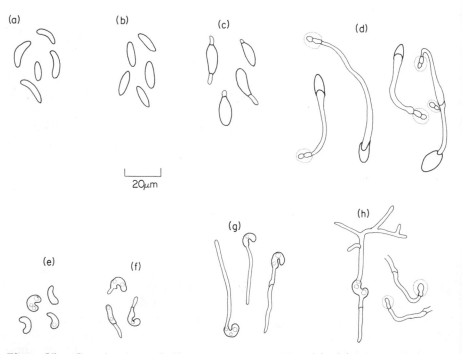

Figure 25 Germination of *Nematoctonus* conidia: (a)—(d) *N. concurrens*, adhesive processes formed on unbranched germ tubes; (e)—(h) *N. haptocladus*, adhesive processes formed after branching of germ tube

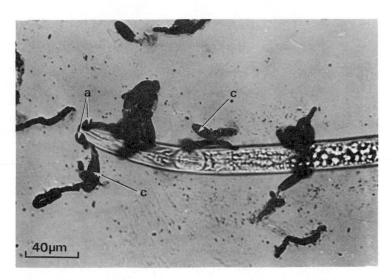

Figure 26 Production of nematotoxic compounds by *Nematoctonus*. Individual of *Aphelenchus avenae* killed after contact with adhesive processes (a) arising from germinating conidia (c) of *N. concurrens*

are not usually themselves adhesive but, on germinating, give rise to adhesive organs either directly upon the spore or on the germling hyphae (Figure 25a–h). Nematodes with such adhering conidia become immobile and die within 24–48 hours of the initial contact, and well before there is any sign of penetration by the fungus. There is then rapid degeneration of the body contents and oil droplets appear within the cadaver (Figure 26). Thermostable, water-soluble compounds have been obtained from cell-free washings of germinating conidia which cause 70–95% mortality in nematode populations within 56 hours of application. These compounds are only produced by conidia when they are germinating and their production is usually, but not invariably, coincident with the appearance of the adhesive glandular organs upon the conidia (Figure 27). Their biochemical nature is unknown but they seem to be of relatively high molecular weight and they may be, in part, polysaccharide. They may therefore be identical to, or closely associated with, components of the adhesive fluid on the glandular organs. The toxins are unspecific in their action and affect a wide range of nematodes including free-living, plant-parasitic and animal-parasitic species. They must be effective in extremely small amounts since contact between a large and vigorous host and a single conidium invariably results in immobilization and death.

The question then arises as to what biological advantages might accrue in nature from the ability of conidia to secrete nematotoxins. In those fungi that infect by adhesive spores, immobilization of the host would reduce the risk of adhering conidia being detached as the nematode moves through soil or other substrata. In relation to this, it is interesting to note that attempts to infect nematodes with zoospores of *Catenaria anguillulae* are much less successful in sand culture than in liquid culture, probably because of the encysted spores being dislodged from their attachment points as the nematodes move through the sand (Sayre and Keeley, 1969). Where toxins are produced, rapid host death

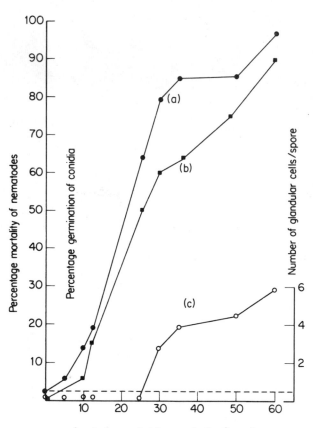

Figure 27   Production of nematoxic compounds by *Nematoctonus tripolitanius*: (a) mortality of *Aphelenchus avenae* after 56 hours exposure to washings from conidia taken at various times over a 60 hour germination period; (b) percentage germin- ation of conidia; (c) average numbers of adhesive cells produced by each conidium. The dotted line indicates mortality level of *A. avenae* after 56 hours exposure to distilled water

and internal disorganization would make readily assimilated nutrients available to the fungus immediately on penetration, at which time reserves in the spore might be severely depleted or exhausted. The host's own enzymes that are bringing about its internal dissolution are thus being indirectly made use of by the fungus. Those endosymbiotic species whose conidia are ingested may not have evolved toxin-secreting mechanisms since their spores become quickly and securely lodged within the host.

Biological Control

Some attempts have been made to control soil-borne plant-parasitic nematode populations under both natural and artificial conditions through the use of endosymbiotic fungi, but the results have not been encouraging (Sayre and Keeley, 1969; Duddington and Wyborn, 1972). Most, if not all, plant-parasitic nematodes have their mouthparts in the form of a stylet so that only adhesive-spored fungi are likely to be potential agents of biological control. Theoretically, addition of sufficient numbers of conidia or of diseased nematodes to soil should result in an epizootic, during which the nematode population would be reduced (Cooke, 1968). Using *Nematoctonus* species it has been shown that this situation is difficult, if not impossible, to bring about.

In order to achieve successful biological control of soil nematode populations through the use of adhesive-spored fungi of the *Nematoctonus* type four criteria must be fulfilled. First, on addition to the soil, conidia must germinate rapidly and adhesive cells must be formed quickly on their germ tubes or germling hyphae. Second, conidia or germlings, or both, must survive in a viable condition for a sufficiently long period to maximize the chance of contact being made with a host. Third, if successful infection and exploitation occur then, after exhaustion of the host's body contents, there must be rapid formation of external fertile hyphae and abundant sporulation. Finally, the conidia so produced must themselves survive and germinate successfully. When *Nematoctonus* conidia are added to non-sterile soil they are susceptible to mycostatic effects and few of them germinate. The germ tubes of those which do germinate become quickly lysed so that there is rapid loss of loci at which hosts might be contacted. In addition, in soil, adhesive process formation on germlings does not always take place (Giuma and Cooke, 1974*b*). In view of these observations it is difficult to see how even massive spore inocula could bring about satisfactory control. Other adhesive-spored fungi may have greater potential, particularly those species whose spores are coated with adhesive at their formation and which do not, therefore, require to germinate before host contact is made.

52

Bibliography

Aschner, M. and Kohn, S., 1958. *J. gen. Microbiol.*, 19: 182—189.
Barron, G. L., 1969. *Can. J. Bot.*, 47: 1899—1902.
Barron, G. L., 1970. *Can. J. Bot.*, 48: 329—331.
Bricklebank, J. and Cooke, R. C., 1969. *Trans. Br. mycol. Soc.*, 52: 347—349.
Capstick, C. K., Twinn, D. and Waid, J. S., 1957. *Nematologica*, 2: 193—201.
Cooke, R. C., 1968. *Phytopathology*, 58: 909—913.
Cooke, R. C. and Godfrey, B. E. S., 1964. *Trans. Br. mycol. Soc.*, 47: 61—75.
Davidson, J. G. N. and Barron, G. L., 1973. *Can. J. Bot.*, 51: 1317—1323.
Drechsler, C., 1935. *Mycologia*, 27: 6—40.
Drechsler, D., 1941a. *Mycologia*, 33: 248—269.
Drechsler, C., 1941b. *Phytopathology* 31: 773—801.
Drechsler, C., 1946a. *Mycologia*, 38: 1—23.
Drechsler, C., 1946b. *Mycologia*, 38: 120—143.
Drechsler, C., 1949. *Mycologia*, 41: 229—251.
Drechsler, C., 1955. *Mycologia*, 47: 364—388.
Drechsler, C., 1959. *Mycologia*, 51: 787—823.
Drechsler, C., 1963. *Am. J. Bot.*, 50: 839—842.
Duddington, C. L. and Wyborn, C. H. E., 1972. *Bot. Rev.*, 38: 545—565.
Giuma, A. Y. and Cooke, R. C., 1971. *Trans. Br. mycol. Soc.*, 56: 89—94.
Giuma, A. Y. and Cooke, R. C., 1972. *Trans. Br. mycol. Soc.*, 59: 213—218.
Giuma, A. Y. and Cooke, R. C., 1974a. *Trans. Br. mycol. Soc.*, 63: 400—403.
Giuma, A. Y. and Cooke, R. C., 1974b. *Soil. Biol. Biochem.*, 6: 217—220.
Giuma, A. Y., Hackett, A. M. and Cooke, R. C., 1973. *Trans. Br. mycol. Soc.*, 60: 49—56.
Karling, J. S., 1944. *Lloydia*, 7: 328—342.
Karling, J. S., 1946. *Lloydia*, 9: 1—12.
Sayre, R. M. and Keeley, L. S., 1969. *Nematologica*, 15: 492—502.
Sorokine, N., 1876. *Annls. Sci. nat. (Bot.), Sér. B*, 4: 62—71.

Chapter 4

Fungi that Attack Insects

Entomogenous fungi may attack the egg, larval or adult stages of insects. Some grow entirely superficially over the host's exterior, living there saprotrophically and deriving their nutrients directly from the cuticle or from host secretions of various kinds. Nevertheless, the host may become deformed and its normal breeding habits be adversely affected by such infections (Madelin, 1963). It is probable that these fungi are normally saprotrophs on dead organic matter but can use the exoskeleton of living insects as an abode when environmental conditions favour this. Some may enter the insect through wounds, but infections are then localized by strong host reactions to invasion (Bucher and Bracken, 1966). The majority of severe or fatal mycoses of insects are, however, caused by antagonistic endosymbiotic fungi. Some of these, for example species of Entomophthorales, are ecologically obligate symbionts while others, in particular hyphomycetous species, are facultative. The latter probably survive in nature as saprotrophs, and some of them live successfully in the absence of host animals or during periods when suitable conditions for infection do not obtain.

At each stage of the association between an endosymbiont and its host a number of physical and physiological problems must be overcome by the fungus. It must be able to enter the insect successfully and, once within living tissues, must be able to grow and become distributed throughout the body. When the host becomes moribund or dies, the fungus must then be able to resist or obviate the effects of putrefying bacteria within the tissues and, finally, sporulate.

1. Entomogenous Hyphomycetes

The most commonly encountered entomogenous Hyphomycetes belong to a relatively few genera, the majority being species of *Beauveria, Metarrhizium* and, less frequently, *Aspergillus, Hirsutella* and *Paecilomyces* (Madelin, 1966). All are facultative, and all can be grown axenically where their nutrient requirements are usually relatively simple, although some species of *Hirsutella* may require organic sources of nitrogen in the form of amino acids, particularly glutamic acid (Maclcod, 1959, 1960).

Infection

Infection is usually effected by conidia and there are four possible routes of entry, directly through the outer integument; via the digestive tract; through tracheae; or through wounds. Successful artificial infection can be produced by introducing spores into fresh wounds, but the natural importance of this route seems to be low. Despite this, a few fungi depend entirely on wound infection for entry (Hurpin and Vago, 1958; Jolly, 1959). Similarly, some species apparently enter certain hosts only through tracheae while infecting others by direct penetration of the cuticle (Madelin, 1963). Insects allowed to feed on spores or forcibly fed with them by means of a micro-injector may become infected, but again whether oral infection commonly occurs in nature is not known. The major entry route for the most important entomogenous Hyphomycetes is through the cuticle. For spore germination to take place on the integument surface a high relative humidity, usually in excess of 80%, is necessary, and in natural populations of insects disease severity is frequently linked with relative humidity levels. It is possible that in less humid conditions germination can still take place if the humidity of the air in the micro-environment of the insect's body surface is above ambient.

The integument presents two major barriers to infection; the outer epicuticle which is typically waxy, and the underlying procuticle which is composed principally of protein and chitin. The most detailed studies of cuticle penetration have been made using *Metarrhizium anisopliae* on the larvae of various wireworm species, but it is probable that other fungi behave towards their hosts in a broadly similar manner (McCauley, Zacharuk and Tinline, 1968; Sannasi, 1969, Zacharuk, 1970a,b).

Germ tubes arise from conidia and, after growing for a short distance, their tips become swollen to form ovoid or elongate appressoria. A single appressorium may proliferate to form clusters of cells making up a complex infection cushion. The appressorium or infection cushion then secretes a mucilaginous sheath which anchors it firmly to the integument. (Figure 28a). Appressorium formation appears to be stimulated by contact of the germ tube tip with the integument since appressoria are formed by *M. anisopliae*, *Beauveria bassiana* and *Paecilomyces farinosus* when their conidia germinate in contact with glass slides or paraffin wax membranes (Madelin, Robinson and Williams, 1967).

In wireworm cuticle the procuticular layer is distinctly laminated through most of its thickness and is traversed by numcrous helical canals containing lipids, these canals being continuous with fine straight channels that cross the epicuticle. These too contain lipid material. Similar substances form a thin layer over the surface of the epicuticle but during the formation of appressoria or infection cushions this fatty layer disappears in the immediate area of contact. This may allow improved adhesion of the fungus to the host and may also supply the fungus with nutrients additional to those contained within the spore.

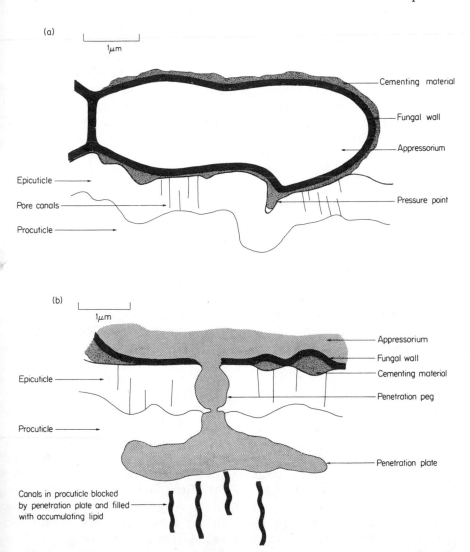

Figure 28   Penetration of wireworm integument by *Metarrhizium anisopliae*: (a) appressorium arising from germ tube of conidium; (b) formation of penetration peg and penetration plate. Diagrams based on a series of electron-micrographs in Zacharuk, 1970 *a,b*

Each appressorial cell forms one to several pressure points on the epicuticle so that the latter becomes indented. There is histolysis of the epicuticle at these pressure points which leads to the production of a cavity below each point. The appressorial wall above each cavity then breaks down and a fine penetration peg grows into it. On passing through the epicuticle and into the procuticle the tip of the peg expands in a direction parallel to the procuticular laminae to form a flattened, discoid penetration plate (Figure 28b). This blocks the procuticular canals so that there is accumulation within them of lipid. There is also complete histolysis of the outer procuticle around the penetration plate.

Lateral penetration hyphae arise from the plate and these in turn give rise to chains or sheets of cells lying among the laminae of the outer procuticle. From these cells, vertical hyphae grow towards the interior of the procuticle for a short distance and then give rise to further chains or sheets of cells. This process is repeated several times so that penetration takes place in a stepwise manner, cells being produced between laminae followed by the formation of hyphae which pass through the laminae. As the inner procuticle is approached the vertical hyphae become longer, and fewer inter-laminar cells are formed. Finally, several hyphae break free of the procuticle and enter the body cavity. During hyphal penetration of the procuticle there is no obvious histolysis, but there is mechanical separation of the laminae and of the cuticular fibrils within these as the fungus passes through or between them.

Ultrastructural studies indicate that penetration of both epicuticle and procuticle is brought about by a combination of physical and enzymatic activity of the fungus (Zacharuk, 1970b). There is never direct movement of the fungus through these layers into the body cavity but always alternate production of masses of cells and hyphae. It is probable that this stepwise process represents the establishment of successive sites from which the fungus can exploit the integument as a nutrient base in order to build up sufficient energy reserves to support further inward growth.

There is abundant evidence that entomogenous Hyphomycetes produce lipases, proteases, and chitinases, the latter being adaptive enzymes in at least some species (Huber, 1958; Gabriel, 1968a). The lipases and proteases act first on the fatty and proteinaceous components of the integument, and chitinolytic activity follows after the action of these has begun. In addition, polysaccharides within the cuticle may be broken down to release galactose and glucose. As well as destroying the physical integrity of the integument, enzyme action releases energy-rich compounds and amino acids for utilization by the fungus during penetration (Leopold and Samšiňáková 1970;

Samšiñáková, Mišíková and Leopold, 1971). Nutrition during the penetration phase is entirely saprotrophic.

Fungus-Host Interactions During Colonization

When hyphae reach the body cavity, stellate or digitate colonies of limited extent are formed which may remain localized for some time (Hurpin and Vago, 1958). The fungus then invades deeper tissues, commonly by abstricting free cells or hyphal fragments from these colonies, but occasionally also by continued growth of unfragmented hyphae (Madelin, 1963; McCauley, Zacharuk and Tinline, 1968). In *M. anisopliae, Beauveria* species, and some other fungi, free cells that can multiply by budding are typical of the early stages of colonization. This yeast-like phase is restricted to the blood, the internal organs not being penetrated. Phagocytes aggregate around and envelop the fungal cells as these float in the haemolymph, but there is no evidence to suggest that free cells are killed by phagocytosis, and in some instances the enveloping host cell is killed by the fungus (Donaubauer, 1959; Kawakami, 1965). Multiplication of fungal cells in the blood may retard or obstruct its circulation but nutrition of the fungus is still primarily saprotrophic, dissolved nutrients being absorbed from the blood.

During later stages of colonization the yeast-like phase gives way to, or becomes associated with, a filamentous phase. The hyphae are at first relatively short but gradually become longer as the disease progresses and have the ability to colonize the host's firm tissues and internal organs. Nutrition then becomes necrotrophic as the fungus first kills host cells and then lives upon them saprotrophically. The kind of dimorphism exhibited here has parallels in some fungi that invade the tissues of vertebrate animals. The yeast habit is presumably a response to one or a number of factors within the insect which preclude normal filamentous growth. As internal conditions change as a result of the activity of free cells within the host, the hyphal phase is initiated.

The inability of free cells to penetrate tissues is probably due to their shape, which does not allow them to become anchored to a surface for a sufficiently long period to permit invasion, rather than to any physiological deficiency (Madelin, 1963). In any particular fungus–insect combination invasion of the organs often proceeds in a definite sequence but this sequence is not identical for all combinations. In general, however, it is the fat body that is attacked first with death of the host occurring rapidly after the onset of tissue invasion. *Aspergillus* species differ from most other Hyphomycetes in that, even in the early stages of colonization, free cells are not produced and there is abundant hyphal growth (Sussman, 1952a; Madelin, 1960). In addition, the thoracic muscles are the first tissues to be invaded, the fat body being for some reason unfavourable to fungal growth.

Entry into, and colonization of, the host by entomogenous Hyphomycetes frequently takes place rapidly. For instance, penetration can take place in less than 24 hours and death may occur within 10 days of this (Madelin, 1966; McCauley, Zacharuk and Tinline, 1968). Very soon after infection, usually within 4—5 days, general behavioural symptoms are developed by the host. At first these are mainly loss of appetite and loss of response to mechanical irritation; later there is increasing lethargy and inability of the insect to right itself if placed on its back. Finally, there is general loss of function and spasmatic movement which leads to paralysis.

Abnormal host behaviour may be solely due to the physical effects of the fungus as it colonizes the blood, so reducing its circulation, or as it forces apart muscles and vital internal organs. However, the fungus might also bring about physiological changes in the host through either histolytic activity or in some other way. When pupae of the cecropia moth *Platysamia cecropia* are infected with conidia of *Aspergillus flavus* there is a rapid increase in host respiration which rises to a peak about a week after inoculation (Sussman, 1952*b*). The maximum respiration rate is approximately ten times that of healthy pupae and the increase mainly reflects an increase in oxygen consumption by host tissues rather than respiratory activity of the colonizing fungus. At the same time, pupae decrease in weight because of increased loss of water through the spiracles due to destruction of tracheoles and ganglionic tissues by the fungus. A number of other changes have been reported for various fungus—host combinations but their significance in host—fungus physiology is not clear. There may be changes in the acidity of the blood and in the activity of certain host enzymes, particularly tyrosinase which is involved in melanization. It has also been suggested that infection may cause endocrine dysfunction (Sannasi and Oliver, 1971).

Although it is probable that no single activity of the fungus is by itself responsible for death of the host, there is evidence that toxins may sometimes make an important contribution. Adult house flies bearing germinating conidia of *Beauveria bassiana* become paralysed before penetration hyphae enter the body (Dresner, 1950). *M. anisopliae* frequently kills its hosts before extensive invasion of vital organs has taken place, while in other fungi limited invasion of the integument, blood, or gut is sufficient to cause paralysis and death (Roberts, 1966*a*). At least some of the toxic compounds produced by *B. bassiana* are proteolytic enzymes, but *M. anisopliae* possesses toxins of relatively low molecular weight (Roberts, 1966*a,b*, Kučera and Samšiňáková, 1968). Two of these, destruxin A and destruxin B, have been characterized and are cyclic depsipeptides composed of five amino acids and a fatty acid in equimolar proportions, the two toxins differing only in the nature of the fatty acid. Injection of them into larvae of the silkworm *Bombyx mori* or the greater waxmoth *Galleria mellonella* rapidly induces tetanic paralysis and death, but if low doses are used then some larvae may be able to detoxify the compounds. Large amounts of destruxins have been detected in the blood of living, infected larvae and this, together with the observation that in pure culture *M. anisopliae* produces destruxins before the lytic phase of its growth, strongly suggests that they have a significance in the natural progression of disease.

Post-Mortem Events

On death of the host the mycelium spreads rapidly to fill the whole body. If the host is a larva, then there is commonly rapid desiccation and hardening of its formerly soft body and water droplets are exuded at the body surface. A compact mass of mycelium is finally formed within the more or less intact integument and this part-fungus—part-animal structure resembles a fungal stroma or sclerotium in both structure and function. The fungus can remain dormant within it for long periods, particularly when conditions are dry. If conditions are moist, and a favourable temperature obtains, then hyphae emerge through the integument and conidiophores bearing numerous conidia are produced. Sporulation can and does take place immediately after host death without the intervention of a dormant phase.

Colonization of a dead host by exogenous bacterial and fungal saprotrophs is a possibility but seems to take place only rarely, and there is also an absence of corruption caused by endogenous putrefying organisms. The activity of alien organisms may be curtailed by antimicrobial compounds produced by the entomogenous fungus. One such compound, oosporein, has been isolated from *B. bassiana* (Vining, Kelleher and Schwarting, 1962).

Host Resistance

Hyphomycetous species can, in general, attack different developmental stages of a wide range of host insects, but particular strains or isolates of some fungi may only infect certain hosts (Madelin, 1963). This implies the operation of host resistance factors, some of which prevent contact of the integument by inoculum while others act after such a contact has been made. For example, the possession of dense hairs or the use of frequent grooming movements may contribute to the prevention of successful contact, but it is post-contact factors that are probably the most important in conferring host resistance.

If an infected larva moults while penetration of its integument is taking place, then it might be expected that the fungus would be shed together with the cast-off moult. It has, however, been shown that *M. anisopliae* can pass from the moult to the new cuticle and is resistant to the histolytic activity of moult enzymes. Hyphal growth anchors the exuvium to the host and so prevents normal ecdysis (Zacharuk, 1973). Under normal circumstances, once the integument has been penetrated infection progresses despite any phagocytic activity in the host's blood, or the laying down of host pseudotissues around free cells or hyphae. It is, therefore, the nature of the integument that determines whether or not progressive infection will take place. The epicuticle is the primary barrier to infection, and mechanical resistance to invasion is probably conferred by its waxy outer layers since artificial de-waxing of the epicuticle increases susceptibility to invasion (Sussman, 1952c). It also contains chemical resistance factors (Koidsumi, 1957; Kawase, 1958; Sannasi and Oliver, 1971). If the epicuticular lipids of larvae of the silkworms *Bombyx mori* and *Chilo simplex* are removed, then the larvae become highly susceptible to infection by *Aspergillus flavus*, the extracted lipids having antifungal properties attributable to saturated fatty acids such as capric and caprylic acid. During infection of the velvet-mite *Dinothrombium giganteum* by *A. flavus*, all regions of the cuticle are normally attacked, but germ tube penetration occurs most rapidly in regions where setae have been shed from their cuticular sockets and cuticular lipids are absent. Integumental phenolic compounds, for example protocatechuic acid, are also fungistatic towards some fungi.

2. The Entomophthorales and Other Fungi

Entomogenous species of the Entomophthorales, members of the Pyrenomycete genus *Cordyceps*, and the Chytridiomycete *Coelomomyces* are all either ecologically or absolutely dependent upon insects. Most appear to be obligate necrotrophs and outside the body of the host they exist only as short-lived or resistant spores or sporangia. It is interesting to note that some *Entomophthora* species will also attack mammalian tissues (Emmons and Bridges, 1961).

The Entomophthorales

Although species in this group may not be absolutely obligately entomogenous, some are difficult to bring into axenic culture, and most seem to have exacting nutrient requirements, growing successfully only on semi-defined media containing peptone or liver extract. However, two common species *Entomophthora apiculata* and *E. coronata* are autotrophic for vitamins and other growth factors and can be grown on defined media. There are three known genera that attack insects, *Entomophthora, Massospora* and *Strongwellsea*, but investigations on the biology of this group have been largely confined to *Entomophthora* species. Species of *Entomophthora* infect their hosts by means of adhesive conidia which become firmly fixed to the integument, their germination being followed by a penetration process the details of which are unknown. The outer cuticular layers become raised during penetration and there is histochemical evidence for the action of lipases, proteinases, and chitinolytic enzymes which may produce this loosening of the endocuticular lamellae (Gabriel, 1968b). Although direct penetration is the normal entry route, conidia have been observed to germinate in the oesophagus and to penetrate its wall. Direct penetration is usually completed in 2–12 hours.

Conidia remain viable for only a short time after their formation, having a maximum life span of about 2 weeks. Successful germination and penetration requires a relative humidity greater than 95%, and some species require levels of 100%. At relative humidities below 90%, viability is lost within 6–9 hours. Germination requirements are thus much more exacting than those of the Hyphomycetes (Yendol, 1968).

Having gained access to the host's body the fungus colonizes it either through filamentous growth, or, more commonly, by the production of large, unicellular hyphal bodies (MacLeod, 1963). These do not circulate in the blood stream, as do the relatively small free cells of the Hyphomycetes, but are restricted to the solid tissues. Conidia germinating on artificial media give rise to naked protoplasts and if the latter are injected into the haemocoel then typical infection occurs. Such protoplasts may, therefore, play a part in the natural colonization process (Tyrrell and MacLeod, 1972). Within 32 hours of infection the fat body and musculature are penetrated (Prasertphon and Tanada, 1968). Death usually occurs after about 80 hours but can take place in as short a time as 12 hours (Yendol and Paschke, 1965).

Behavioural symptoms resulting from infection resemble those caused by hyphomycetous fungi, and it is probable that production of lytic enzymes, in particular large quantities of proteases, play an important part in killing the host (Jönsson, 1968). It is also possible that toxins are involved (Prasertphon, 1967; Yendol, Miller and Behnke, 1968). Some *Entomophthora* species, for example *E. apiculata* and *E. coronata*, if grown in liquid media release compounds which, when injected, will kill insect larvae. Blackening of the blood occurs within 3 hours and death 1–2 days later. Low doses may not be lethal but metamorphosis is inhibited. There is some evidence that the toxins may have some kind of specific action, since those obtained from *E. virulenta* act against larvae of the face fly *Musca autumnalis* but not against those of *Galleria mellonella*. The compounds are heat labile, becoming rapidly inactivated even at room temperature, and appear to be proteins. It has been suggested that they do not act directly but induce toxin production by host tissues (Yendol, 1968). There is also the possibility that they are, after all, proteinases, but this, and their role in the natural progress of disease, remains to be determined. After death of the host vegetative development within it ceases. Sometimes hyphae grow through the host's ventral tissues and anchor its body firmly to a surface of some kind. Hyphal bodies produce stout conidiophores which break through the integument and then give rise to conidia. Sexual reproduction within the cadaver results in the production of resistant zygospores.

The infection process of *Massospora* species has not been observed in detail, but after infection hyphal bodies do not appear until host tissues are in an advanced state of disintegration. Hyphal bodies then line cavities within the tissues and give rise to short conidiophores bearing conidia (Speare, 1921; Goldstein, 1929). *Massospora* infections do not kill the host rapidly and vegetative growth of the fungus is confined to the softer tissues of the posterior segments of the insect's body. As each segment becomes colonized, and then filled with spores, it is sloughed away so that finally the host, though still alive and mobile, may possess only head and thorax. *Strongwellsia castrans*, the single representative of this genus, also produces spore-filled cavities within its host and each cavity is finally connected to the exterior by means of a small hole in the abdomen (Nair and McEwen, 1973). This little-known fungus is worthy of note in that infections are non-lethal, the only apparent effect being suppression of gonad development (Batko and Weiser, 1965).

Cordyceps

Cordyceps species have been described from most insect orders, and also from spiders, while some attack other fungi (McEwen 1963). The limited evidence available indicates that most are host-specific, although a single host species may be susceptible to more than one species of *Cordyceps*. Some have been grown on natural or semi-defined media but their precise nutrient requirements are not known. It is of interest that the conidial states of a number of species are within the entomogenous Hyphomycete genera *Hirsutella* and *Isaria*.

Details of penetration and colonization are almost totally lacking but, since no signs of the fungus have been found in the gut during the primary stages of infection, entry is presumed to be effected by ascospore germ tubes passing directly through the cuticle. It is known that they can produce chitinolytic enzymes (Mathieson, 1949; Huber, 1958). Hyphal bodies quickly appear in the haemolymph and these initially cluster on the surfaces of organs bathed by the blood. Later

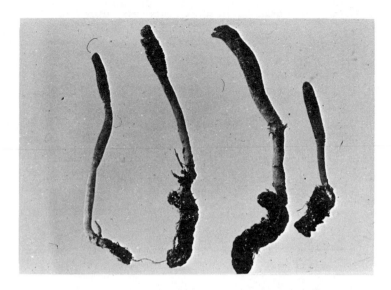

Figure 29   *Cordyceps militaris*, stromata arising from the dead
bodies of unidentified insect larvae

they occur in the muscles and fat body where, as infection proceeds,
they are replaced by chains of cells. After death the host shrinks and
dehydrates. A conspicuous clavate or cylindrical stroma eventually
arises from it and ascospores are ejected from perithecia submerged
within this (Figure 29). The tissues of dead hosts are resistant to decay
and this seems to be due to production by the fungus of cordycepin, an
antimicrobial compound. This is an adenine nucleoside which inhibits
nucleic acid synthesis, but its possible effect on living hosts has not yet
been established (Rottman and Guarino, 1964).

Coelomomyces

Coelomomyces species are normally found within the larvae of
mosquitos and while some are restricted to a single host species others
have a wide host range. The fungus is confined to the coelomic cavity
within which it produces branched coenocytic 'hyphae' (Umphlett,
1962, 1964). These are remarkable for their lack of a true cell wall.
Even more remarkable is the ability of the outer boundary of the
'hyphae' to dissolve rapidly in water if they are removed from the host.
Hyphal bodies are produced, presumably by budding, and these
circulate in the haemolymph, passing easily through the heart. The
fungus exploits adipose tissue and infected larvae normally die before

pupation. *C. psorophorae* is heteroecious, its alternate host being
Cyclops vernalis (Whisler, Zebold and Shcmanchuk, 1974).

Termitaria snyderi

One of the few entomogenous fungi that is undoubtedly biotrophic is
the Hyphomycete *Termitaria snyderi*, which lives upon the exoskeleton
of termites (Kahn and Aldrich, 1975). Hyphae penetrate the cuticle,
apparently by enzyme secretion, and haustoria are then formed within
the insect (Figure 30). The bounding membranes of the host's
epidermal cells are distorted, but not penetrated, and these cells remain
alive. A unique feature of the haustorium is the presence of

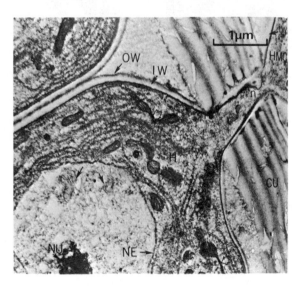

Figure 30 *Termitaria snyderi*, penetration of the
subterranean termite *Reticulotermes*. Electronmicrograph of median section through the neck of
the haustorium. The smoothness of the hole in the
host cuticle (CU) through which the neck (n)
passes suggests that penetration is enzymatic.
There is continuity of the inner layer (iw) of the
haustorial mother cell wall (HMC) with that of the
inner layer (IW) of the haustorial wall through the
neck. The haustorium has a large nucleus (NU)
bounded by a double-membraned envelope (NE).
Memaranous structures within the nucleus are
arrowed and open arrows indicate constricted
fungal mitochondria. From Khan and Aldrich,
1975; by permission of *Journal of Invertebrate
Pathology*

minimicrotubules within its cytoplasm. These organelles have not been observed in either other fungi or other organisms.

3. Biological Control of Insect Populations

Recently there has been increased emphasis on investigating the use of insect-destroying fungi as a complete or partial alternative to chemical insecticides (Steinhaus, 1957; Tanada, 1959, 1963; Hall, 1963; Weiser, 1970). Two broad, opposing groups of factors influence insect populations and these are always to a greater or lesser extent counterbalanced. There is first the biotic potential of the population, that is its inherent ability to increase or survive. Second there is environmental resistance, which comprises all physical and biological factors that tend to reduce biotic potential. In certain circumstances entomogenous fungi may be an important component of environmental resistance and, if their contribution could be increased in some way, effective biological control would be a possibility. Natural epizootics frequently occur, and are of some significance in natural population regulation, so that the way seems open to artificial regulation methods. In general, epizootics can be expected to occur naturally, or to be successfully induced, mainly in conditions of high host density, but they might also occur at low density. In the latter situation a population reduction would take place only if the fungus were dispersed widely enough. Unfortunately, natural epizootics often occur too late in the growing season to prevent the maximum insect population from being reached (Dean and Wilding, 1971).

Fungi for biological control of insects should ideally possess four attributes; a good infection capability, high post-infection virulence, adequate survival potential, and an efficient dispersal mechanism. In addition, they should be used against that developmental stage of the insect which is most suitable for infection, colonization and sporulation. Hyphomycetes, *Entomophthora* species, and *Coelomomyces* species have all at some time been used in small- and large-scale control experiments. Hyphomycete spores have been used directly as insecticides with or without sublethal doses of conventional chemicals. Results have been variable, the spore doses required are possibly too high for economic application, and their effects are non-persistent (Clarke, Kellen, Fukuda and Lindegren, 1968; Bell and Hamalle, 1970). Lack of success is often due to uncontrollable environmental conditions that are unfavourable to the spread or development of the fungi. Recently, *Coelomomyces* species have been tested for their ability to control natural mosquito populations. Although the fungi can be established in mosquito populations which are normally free from them, their long-term effects upon the numbers of larvae surviving remains to be established (Umphlett, 1970).

66

Bibliography

Batko, A. and Weiser, J., 1965. *J. Invert. Path.*, 7: 455—463.
Bell, J. V. and Hamalle, R. J., 1970. *J. Invert. Path.*, 15: 447—450.
Bucher, G. E. and Bracken, G. K., 1966. *J. Invert. Path.*, 8: 193—204.
Clarke, T. B., Kellen, W. R., Fukuda, T. and Lindegren, J. E., 1968. *J. Invert. Path.*, 11: 1—7.
Dean, G. J.W. and Wilding, N., 1971. *J. Invert. Path.*, 18: 169—176.
Donaubauer, E., 1959. *Sydowia*, 13: 183—222.
Dresner, E., 1950. *J.N.Y. Entomol. Soc.*, 58: 269—278.
Emmons, C. W. and Bridges, C. H., 1961. *Mycologia*, 53: 307—312.
Gabriel, B. P., 1968a. *J. Invert. Path.*, 11: 70—81.
Gabriel, B. P., 1968b. *J. Invert. Path.*, 11: 82—89.
Goldstein, B., 1929. *Am. J. Bot.*, 16: 394—401.
Hall, I. M., 1963. In E. A. Steinhaus (ed.), *Insect Pathology*, Vol. 2, p. 477—517, Academic Press, New York.
Huber, J., 1958. *Arch. Mikrobiol.*, 29: 257—276.
Hurpin, B. and Vago, C., 1958. *Entomophaga.*, 3: 285—330.
Jolly, M. S., 1959. *Annls. Inst. natn. Rech. agron., Paris. sér. C.*, 10: 37—43.
Jönsson, A. G. 1968. *Appl. Microbiol.*, 16: 450—457.
Kawakami, K., 1965. *J. Invert. Path.*, 7: 203—208.
Kawase, S., 1958. *Nature, Lond.*, 181: 1350—1351.
Khan, S. R. and Aldrich, H. C., 1975. *J. Invert. Path.*, 25: 247—260.
Koidsumi, K., 1957. *J. Insect Physiol.*, 1: 40—51.
Kučera, M. and Samšiňáková A., 1968. *J. Invert. Path.*, 12: 316—320.
Leopold, J. and Samšiňáková A., 1970. *J. Invert. Path.*, 15: 34—42.
McCauley, V. J. E., Zacharuk, R. Y. and Tinline, R. D., 1968. *J. Invert. Path.*, 12: 444—459.
McEwen, F. L., 1963. In E. A. Steinhaus (ed.) *Insect Pathology*, Vol. 2, p. 273—290, Academic Press, New York.
MacLeod, D. M., 1959. *Can. J. Bot.*, 37: 819—834.
MacLeod, D. M., 1960. *J. Insect. Path.*, 2: 139—146.
MacLeod, D. M., 1963. In E. A. Steinhaus (ed.). *Insect Pathology*, Vol. 2, p. 189—231, Academic Press, New York.
Madelin, M. F., 1960 *Endeavour*, 19: 181—190.
Madelin, M.,F., 1963. In E. A. Steinhaus (ed.), *Insect Pathology*, Vol. 2, p. 233—271, Academic Press, New York.
Madelin, M. F., 1966. *A. Rev. Ent.*, 11: 423—448.
Madelin, M. F., Robinson, R. K. and Williams, R. J., 1967. *J. Invert. Path.*, 10: 404—412.
Mathieson, J., 1949. *Trans. Br. mycol. Soc.*, 32: 113—116.
Nair, K. St. and McEwen, F. L., 1973. *J. Invert. Path.*, 22: 442—449.
Prasertphon, S., 1967. *J. Invert. Path.*, 9: 281—282.
Prasertphon, S. and Tanada, Y., 1968. *J. Invert. Path.*, 11: 260—280.
Roberts, D. W., 1966a. *J. Invert. Path.*, 8: 212—221.
Roberts, D. W., 1966b. *J. Invert. Path.*, 8: 222—227.
Roberts, D. W., 1969. *J. Invert. Path.*, 14: 82—88.
Rottman, F. and Guarino, A., 1964. *Biochim. biophys. Acta.*, 80: 640—647.
Samšiňáková, A., Mišiková, S. and Leopold, J., 1971. *J. Invert. Path.*, 18: 322—330.
Sannasi, A., 1969. *J. Invert. Path.*, 13: 4—10.
Sannasi, A. and Oliver, J. H., 1971. *J. Invert. Path.*, 17: 354—365.
Speare, A. T., 1921. *Mycologia*, 13: 72—82.
Steinhaus, E. A., 1957. *A. Rev. Microbiol.*, 11: 165—182.
Sussman, A. S., 1952a. *Ann. ent. Soc. Am.*, 45: 233—245.

Sussman, A. S., 1952b. *Mycologia*, **44**: 493—505.

Sussman, A. S., 1952c. *Ann. ent. Soc. Am.*, **45**: 638—644.

Tanada, Y., 1959. *A. Rev. Ent.*, **4**: 277—302.

Tanada, Y., 1963. In E. A. Steinhaus (ed.), *Insect Pathology*, Vol. 2, p. 423—475, Academic Press, New York.

Tyrell, D. and MacLeod, D. M., 1972. *J. Invert. Path.*, **19**: 354—360.

Umphlett, C. J., 1962. *Mycologia*, **54**: 540—544.

Umphlett, C. J., 1964. *Mycologia*, **56**: 488—497.

Umphlett, C. J., 1970. *J. Invert. Path.*, **15**: 299—305.

Vining, L. C., Kelleher, W. J. and Schwarting, A. E. 1962. *Can. J. Microbiol.*, **8**: 931—933.

Weiser, J. 1970. *A. Rev. Ent.*, **15**: 245—256.

Whisler, H. C., Zebold, S. L. and Shemanchuk, J. A. 1974. *Nature, Lond.*, **251**: 715—716.

Yendol, W. G. 1968. *J. Invert. Path.*, **10**: 116—121.

Yendol, W. G., Miller, E. M. and Behnke, C. N. 1968. *J. Invert. Path.*, **10**: 313—319.

Yendol, W. G. and Paschke, J. D. 1965. *J. Invert. Path.*, **7**: 414—422.

Zacharuk, R. Y. 1970a. *J. Invert. Path.*, **15**: 81—91.

Zacharuk, R. Y. 1970b. *J. Invert. Path.*, **15**: 372—396.

Zacharuk, R. Y. 1973. *J. Invert. Path.*, **21**: 101—106.

Chapter 5

Fungi of Larger Invertebrates and Vertebrates

Fungi commonly cause diseases of larger invertebrates, for example crabs, crayfish, and lobsters (Johnson, 1968; Unestam, 1973). More importantly they are also found on and within vertebrates, particularly fish, birds, and mammals. Fungi of crustacea and fish are widespread and may sometimes be of economic importance, as are some fungi of edible molluscs, but the kinds of fungi involved are very diverse, they have a great range of effects upon their hosts, and details of host—fungus relationships are generally unknown. (Carmichael, 1966; Stuart and Fuller, 1968; Sindermann, 1970). Thus, at the moment, it is impossible to make a comprehensive treatment of them within the context of symbiology, and they will not be considered here.

There is a large and rapidly growing literature concerned with the fungi of warm-blooded animals and in particular those fungi that cause disease. Medical and veterinary mycology is assuming an increasing importance since the occurrence of such infections, especially of man, is for various reasons becoming more frequent, and because domestic and wild animals may act as reservoirs for diseases of medical importance (Moss and McQuown, 1960; Conant, 1962; Emmons, Binford and Utz, 1970). Quite properly, the greater part of research upon these fungi has taken place in those areas directly related to the clinical diagnosis and treatment of disease (Ainsworth and Austwick, 1959; Emmons, 1960a; Winner and Hurley, 1964). The very breadth of these studies, together with their mainly clinical bias, has necessitated a degree of selection in considering those topics to be discussed here. Only the more fundamental mycological aspects of mammal—fungus relationships are outlined, with particular respect to those involving man.

Many unusual diseases of clinical rarity that are caused by fungi have been recorded in man, and the fungi involved are normally common free-living saprotrophs. It is clear that a great many fungi, provided that they can gain access to either superficial or deep mammalian tissues, do not find the body a *milieu* inimical to their growth (Emmons, 1960b). The well defined, and more commonly encountered, diseases of man

Table 3

Fungi causing diseases of man grouped according to possession of a free-living phase and to their effects on the body

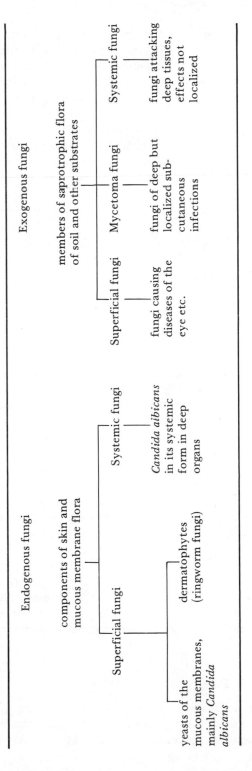

Endogenous fungi

components of skin and mucous membrane flora

Superficial fungi

yeasts of the mucous membranes, mainly *Candida albicans*

dermatophytes (ringworm fungi)

Systemic fungi

Candida albicans in its systemic form in deep organs

Exogenous fungi

members of saprotrophic flora of soil and other substrates

Superficial fungi

fungi causing diseases of the eye etc.

Mycetoma fungi

fungi of deep but localized sub-cutaneous infections

Systemic fungi

fungi attacking deep tissues, effects not localized

number about twenty, and the fungi which cause them belong to a similar number of genera comprising about seventy or eighty species. Most are Hyphomycetes, together with a few Zygomycotina, Ascomycotina and Basidiomycotina. With a few exceptions the production of disease symptoms is due to necrotrophic activity within host tissues, the majority of the fungi not having developed specialized mechanisms for exploiting their hosts (Austwick, 1972).

There are two possible modes of growth *in vivo*, either through normal hyphal extension or through the budding of yeast-like cells. Hyphal growth leads to slow extension through tissues with the maintenance of a compact mycelial colony, while division of yeast-like cells allows rapid dissemination of the fungus through the body of the host. Spread frequently occurs via the blood stream after phagocytosis of the fungal cells, the phagocytes subsequently rupturing to release viable fungal units. Many fungi that grow in a yeast form in tissue produce a filamentous mycelium when growing in axenic culture. The possession of a hyphal or a unicellular growth form has little or no relationship to the pathogenicity of most fungi. The *in vivo* yeast form merely indicates that the environment precludes hyphal growth, one of the many factors responsible being the relatively high temperature of the mammalian body (Scherr and Weaver, 1953; Romano, 1966).

In general the fungi of mammals fall into one of two categories, being either endogenous or exogenous species (Ajello, 1968*a*). Endogenous fungi are those that are considered not to exist normally in a free-living saprotrophic state in nature but to form part of the flora of the body, living either on the skin or mucous membranes. Exogenous fungi are those that are known, or presumed, to exist in a free-living saprotrophic form in nature. These two major categories may be subdivided on the basis of the kinds of disease which their fungi can cause (Table 3). On a medical basis it is usual to place the diseases in one of three major categories and not to emphasize the endogenous or exogenous origin of the causal organism. These categories are *dermatophytoses*, infections limited to the skin and its appendages; *mycetomas*, localized subcutaneous infections causing suppuration and swelling; and *mycoses*, which are systemic infections. On a mycological basis it appears to be more logical to place emphasis on the endogenous or exogenous nature of the fungi, since there are quite distinct differences in the biological characteristics of the members of these two groups.

1. Exogenous Fungi

These enter the body through wounds or when spores are inhaled and proceed to colonize host tissues. Symptoms may vary from the benign to the chronic or severe (Emmons, 1960*b*; Ajello, 1968*a*; Emmons,

Binford and Utz, 1970). Many species are common in soil in areas where the diseases which they cause are endemic, while some are abundant within, and restricted to, animal droppings, particularly those of birds and bats (Emmons, 1954; Ajello, 1962). They have been described as being 'opportunistic' fungi in that, while having a free-living saprotrophic existence, they can, when a suitable opportunity arises, transfer their abode to animal tissues. They are thus basically saprotrophs out of place, as are many of those fungi associated with the more rare fungal diseases of man. This opportunism is reflected in their great taxonomic diversity (Table 4).

Most body tissues and all the major organs, can be colonized necrotrophically by one or other of the known exogenous fungi. The severity of most disease symptoms and the frequently very rapid spread of the fungus through tissues suggests a general lack of compatibility between fungus and host. That the host is not the primary abode of exogenous fungi is also indicated by the observation that natural host to host transmission of them is almost unknown, infection being always through inoculum derived from soil or other substrates. In many respects they behave in a similar manner to the facultative necrotrophic fungi which cause diseases of higher plants. There is one exception that is worthy of note, the Ascomycotina species *Histoplasma capsulatum* which causes histoplasmosis in man.

H. capsulatum can survive and multiply within living host cells. Although some other exogenous species, in particular the yeast *Cryptococcus neoformans*, may become engulfed by host cells and survive, there is little evidence for their multiplication within them. Infection by *H. capsulatum* takes place when conidia are inhaled in dust arising from soil contaminated with bat or bird dung. It can apparently only survive as a saprotroph in soil enriched with these kinds of excreta. Bats may become infected but birds normally do not, possibly because of their relatively high body temperature. The establishment of the fungus in the human body produces symptoms the severity of which vary depending upon whether the fungus remains relatively localized or becomes disseminated (Silverman, Schwarz and Lahey, 1955). The fungus grows in a yeast-like form within reticulo-endothetlial cells so that the mucous membranes, skin, bone marrow, lymphatic system, lungs, spleen, liver, adrenals, kidneys, and central nervous system are all susceptible to colonization. Hyphal growth is produced in axenic culture, although the yeast phase can be induced if the fungus is grown on rich media at body temperature. Branched hyphae have sometimes been observed in human tissues.

Infected host cells multiply rapidly to produce large cell masses which crowd and replace healthy tissues. Cells may be so filled with fungal material that their cytoplasm is not visible, although their nuclei retain their normal appearance and staining reaction. Infected host cells

Table 4
Some genera of exogenous fungi causing diseases in man

Species	Tissues involved	Habitat when free-living
Zygomycotina		
Coccidioides (?)	Respiratory tract, lungs, viscera, bones	Desert soils
Absidia, Cunninghamella,	Subcutaneous tissues, nasal sinuses,	Decaying organic matter
Mucor Rhizopus	Pulmonary and gastric tissues	Dung, rotting fruit
Basidiobolus Entomophthora	Subcutaneous tissues, pulmonary and gastric tissues	Decaying organic matter, soil
Ascomycotina		
Histoplasma	Lymphatic system, lungs, spleen, liver, kidney, central nervous system	Bat and bird dung, soil
Neurospora Gibberella	Cornea	Soil
Hyphomycetes		
Aspergillus	Lungs and other organs, nasal sinuses	Decaying organic matter
Cephalosporium Fusarium Cylindrocarpon Volutella Penicillium Curvularia	Cornea	Soil
Cladosporium Phialophora Scopulariopsis Cercospora Leptosphaeria	Skin and subcutaneous tissues	Soil and wood
Geotrichum	Mucosal surfaces	Fruit, soil, dairy products
Basidiomycotina		
Cryptococcus (Leucosporidium)	Pulmonary and other tissues	Pigeon droppings

eventually rupture to release the enclosed yeast cells. Intracellular growth of *H. capsulatum* will take place in cultures of mammalian histiocytes and also within histiocytes of poikilothermic animals such as frogs and fish (Larsh and Shepard, 1958; Howard, 1959, 1964, 1965, 1967). In mouse histiocytes *H. capsulatum* cells have a generation time of about 11 hours at 37°C, which is comparable with times for axenic cultures of 7 and 11 hours on agar and in liquid media respectively.

When *H. capsulatum* grows and multiplies within host cells the fungus is released from them after about 3 days. During this period there are a number of ways in which the dividing yeast cells could be obtaining their nutrients (Howard and Otto, 1969). They might utilize endogenous reserves contained within those fungal cells responsible for initial infection. However, the short intracellular generation time of the fungus, equivalent to that in axenic culture, precludes this possibility. In addition, starvation of *H. capsulatum* cells before inoculation into histiocyte culture does not affect their subsequent generation time. Alternatively, they might be exploiting nutrients engulfed by the histiocytes at the time at which the infecting cells are ingested. This is unlikely since fungal cells incorporate label from tritiated leucine added to tissue cultures *after* the phagocytic event. Curiously, it appears that this incorporation takes place directly from the culture medium and not from intracellular pools within the histiocyte cytoplasm. This implies that, although the fungus is not adversely affected by its residence within host cells and does not immediately kill the latter, nutrition of the fungus within them is not biotrophic but is saprotrophic. Nutrients diffuse through the host cytoplasm from the medium and are then directly utilized by the yeast cells. This might explain why *H. capsulatum* can multiply within cultures of host cells from such distantly related animals as mammals, amphibia and fish. It is in any case, perhaps, inconceivable that a single fungal cell could contain all the delicate physical and physiological mechanisms necessary to allow it to exploit, biotrophically and with equal success, cells from such widely different animals. Nor, since in nature *H. capsulatum* must rarely come into contact with either frogs or fish, could this species be expected to have evolved such mechanisms. It is more likely that host cells from such different sources tolerate the presence of the fungus without supplying its nutrients directly. This apparent uniqueness of *H. capsulatum* is worthy of further physiological investigation at the cellular level.

2. Endogenous Fungi

The great majority of fungal diseases of man are caused by endogenous fungi but, unlike those caused by exogenous fungi, these diseases are rarely very severe, and are in general limited to the skin and its appendages or to mucous membranes. Endogenous fungi fall into two groups, those causing dermatophytoses and related diseases (ringworm fungi), and a much smaller group comprising mucosal yeasts, the principal species being *Candida albicans*.

Dermatophytes

In contrast to exogenous fungi, these form a taxonomically uniform group consisting mainly of species of two closely related genera of the Hyphomycetes, *Microsporum* and *Trichophyton* (Ajello, 1968b). Where Ascomycotina perfect states are known these belong within the Gymnoascaceae in the Plectomycetes. *Histoplasma capsulatum* has also been erroneously reported as having this kind of perfect state. (Ajello and Cheng, 1967; Kwon-Chung, 1967). Many *Microsporum* and *Trichophyton* species that are not dermatophytes are commonly found free-living on keratin-rich substrates, for example feathers, or in keratin-enriched soil. Although such free-living species are congeneric with dermatophytes, there is good reason to believe that the latter are not established members of the soil microflora despite the occasional reported isolation of them from soil. This view is supported by the observation that transmission of dermatophytes from one host to another is commonly by direct body contact or, if by indirect contact, through the agency of infected skin or hair fragments. For only two species, *M. filum* and *M. gypseum*, is there evidence for infection by contact with contaminated soil, and at least the latter of the two species may be a strictly exogenous fungus (Ajello, 1962; Emmons, Binford and Utz, 1970).

Although deeper tissues may occasionally be invaded it is normally the hair, skin and nails that are colonized, the fungi growing saprotrophically within them. Hair and hair follicles may be destroyed and the epidermis becomes raised, peeling when wet and flaking when dry. Pathological changes such as spongiosis, dermo—epidermal splitting, and inflammation, may be produced in the deeper layers of the epidermis and dermis, from which the fungus is absent, by means of proteases or other factors diffusing from the infection site (Minocha, Pasricha, Mohapatra and Kandhari, 1972).

The well-developed ability of dermatophytes to rapidly destroy keratinized material has aroused much interest. A major assumption which has commonly been made is that dermatophytes can digest keratin and utilize its breakdown products for growth. It has, however, been suggested that only nutrients contained within the non-fibrillar matrix of hair or skin are utilized, and it has been shown that *Trichophyton* species will grow on aqueous extracts from hair and nails (Raubitschek, 1961). During exploitation of these matrices the integrity of the material is physically disrupted by hyphal expansion.

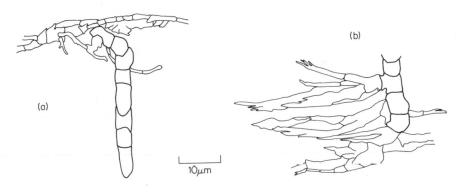

Figure 31 *In vitro* colonization of human hair by *Nannizia persicolor*: (a) frond hyphae on hair surface giving rise to perforating hypha; (b) lateral production of eroding hyphae on lower part of perforating hypha. Redrawn from English, 1967; by permission of *Sabouraudia*

Non-keratinophilic fungi also seem to be capable of growing within hair by using routes afforded by its non-fibrillar components (English, 1965).

Hair destruction by dermatophytes *in vitro* commonly begins with the formation of external frond-like eroding hyphae from which arise, at right angles to them, columnar perforating hyphae (English, 1967). After penetrating the hair for some distance, further lateral eroding hyphae with pointed tips then arise from the perforating hyphae (Figure 31a, b). Electron microscope studies show that during this process there is digestion of the fibrillar keratinous hair material, although not all dermatophytes may do this to the same degree (Baxter and Mann, 1969). In a species such as *T. mentagrophytes* which does digest keratin, the penetrating cells are rich in mitochondria, while in species where digestion is poor, for example *T. ajelloi*, there are many fewer mitochondria. This may reflect a difference in their capability to produce keratinolytic enzymes.

Accepting that keratin degestion does take place in many dermatophytes, the mechanism of keratinolysis still remains to be fully elucidated. An extracellular protease with keratinolytic properties has been isolated from *T. mentagrophytes* and has been purified, but it is capable of dissolving only about 15% of native keratin (Yu, Harmon and Blank, 1968, 1969). It is possible, therefore, that the fungi first denature the keratin in some way and so make it more susceptible to breakdown by less specific proteases. The basis of keratin resistance depends upon cystine bonds within protein-incorporated cystine. Reduction of these bonds may be a key reaction in keratinolysis and appears not to be enzyme-mediated (Kunert, 1968, 1972). The keratin which is denatured in this way can then be acted on by extracellular proteases.

Mention has been made of invasion of deeper tissues by dermatophytes. Deep invasion seems to be associated with the ability of some species to grow in a yeast form (Rippon and Scherr, 1959). Factors bringing about this change in the superficial body layers have not been determined, but such changes can be induced in fungal cells by placing them in the peritoneal cavity of test animals or, *in vitro*, by growing the fungi on cysteine-enriched media. The yeast cells can presumably undergo phagocytosis and then be delivered by the blood stream to deeper tissues or organs, particularly those of the liver and spleen.

Dermatophytes grow well in axenic culture on relatively simple media although they may require vitamins and amino nitrogen (MacKinnon and Artagaveytia-Allende, 1948). While *in vivo* the keratinized body layers are able to fulfil all the nutrient requirements of dermatophytes, there seems to be no nutritional basis for their normally endogenous distribution, their needs being similar to those of the closely related, free-living keratinophilic fungi that cannot or do not invade body tissues. Dermatophytes may lack the ability to compete with other micro-organisms in substrates such as soil even when these are enriched with keratin. Their natural association with man and other animals may be due to their being able to tolerate the normally fungistatic effects of fatty materials in skin and hair (Baxter and Trotter, 1969; McGinnis and Hilger, 1972).

The dermatophytes form a compact group of antagonistic symbionts and most species are obligately dependent on a host for active growth, if not for survival. Their saprotrophic activity in non-living tissues indicates how closely they resemble free-living forms. The pathological effects which they produce seem to be incidental to this saprotrophic existence and are not necessarily caused by factors specifically produced by the fungi for that purpose. In being able to occasionally invade and become necrotrophic in deep-seated tissues they merely share a common capability with a wide range of exogenous fungi. If the latter can be succinctly described as being 'opportunistic' then dermatophytes might equally well be called 'reluctant' antagonists. They are perhaps only a short step away from being obligate neutrals.

Mucosal Yeasts

Many yeasts or yeast-like fungi commonly occur on the skin or mucous membranes where they exist as saprotrophs. Under certain conditions they may become necrotrophic and antagonistic towards the host and cause a variety of diseases that may be either superficial or deep seated. The most frequently encountered fungi of this kind belong to the genus *Candida* and one species, *Candida albicans*, is ubiquitous on man and other animals (Winner and Hurley, 1964; Austwick, Pepin, Thompson and Yarrow, 1966). Under normal conditions it appears to be an obligate neutral since it has rarely been isolated from soil or other non-animal substrates, and there is no evidence that it can grow successfully in such habitats. It may be significant in this respect that inocula of *C. albicans* are transmitted from animal to animal and are not derived from other sources.

In man *C. albicans* is frequent on the membranes of the mouth, digestive tract and vagina, and it also lives as an unobtrusive saprotroph on the body surface. Normally penetration of even the superficial layers of the body does not take place, but a change in the physiological status of the host may result in invasion. Invasion gives rise to a great variety of diseases grouped under the general term of candidoses (Winner, 1966). Predisposition of the body to invasion may be brought about by a number of natural and unnatural events such as pregnancy, endocrine disorders, antibiotic and steroid therapy, or the use of immunosuppressive drugs (Jennison, 1966; Rifkind, Marchioro, Schnek and Hill, 1967).

Candidoses may be superficial, that is restricted to the body surface, or involve deep tissues, all organs of the body being susceptible to invasion. Saprotrophic growth is yeast-like but when invasion occurs this is usually effected by production of hyphae (Gresham and Burns, 1960; Gresham and Whittle, 1961). The conditions required to bring about this transformation *in vivo* are not known but, *in vitro*, transformation to hyphal growth can be induced by provision of exogenous cysteine. Hyphae are obviously more capable of tissue penetration than are unicells since they can bring large mechanical forces to bear, but there may also be physiological differences between the two phases which either impose the neutral state in the case of the yeast phase or allow invasion to take place by the hyphal phase. However, it has been shown that deep invasion can still take place *in vivo* when filamentous growth is impaired by inhibitors, and that there may be no correlation between invasive ability of an isolate and an ability for filamentous growth (Mackenzie, 1964). The yeast cells of *C. albicans* can survive, but perhaps not grow, within living epithelial cells or phagocytes but usually kill them and break out from them in a filamentous form (Louria and Brayton, 1964).

In axenic culture *C. albicans* can use keratin as a sole nitrogen source and there is evidence from electron microscopy that keratin destruction takes place in candidosis lesions (Kapica and Blank, 1957; Montes and Wilborn, 1968). Serum proteins are not utilized by the fungus *in vitro* but their constituent amino acids are taken up. This has led to the suggestion that some of the symptoms of superficial candidoses might be caused by the fungus depriving the host of keratin precursors and so interfering with keratinization (Taschdjian and Kozinn, 1961).

78

Bibliography

Ainsworth, G. C. and Austwick, P. K. C., 1959. *Fungal Diseases of Animals*, Commonwealth Agricultural Bureau.

Ajello, L., 1962. In G. Dalldorf, (ed.), *Fungi and Fungous Diseases*, p. 69—83, Thomas, Illinois.

Ajello, L., 1968a. In G. E. W. Wolstenholme and R. Porter (eds.), *Systemic Mycoses*, p. 130—139, Churchill, London.

Ajello, L., 1968b. *Sabouraudia*, 6: 147—159.

Ajello, L. and Cheng, S-L., 1967. *Mycologia*, 59: 689—697.

Austwick, P. K. C., 1972. In H. Smith and J. H. Pearce (eds.), *Microbial Pathogenicity in Man and Animals*, p. 251—268, 22nd. *Symp. Soc. gen. Microbiol*, Cambridge University Press.

Austwick, P. K. C., Pepin, G. A., Thompson, J. C. and Yarrow, D., 1966. In H. I. Winner and R. Hurley (eds.), *Symposium on Candida Infections*, p. 89—99, Livingstone, Edinburgh and London.

Baxter, M. and Mann, P. R., 1969. *Sabouraudia*, 7: 33—37.

Baxter, M. and Trotter, M. D., 1969. *Sabouraudia*, 7: 199—206.

Carmichael, J. W., 1966. *Sabouraudia*, 5: 120—123.

Conant, N. F., 1962. In G. Dalldorf (ed.), *Fungi and Fungous Diseases*, p. 3—10, Thomas, Illinois.

Emmons, C. W., 1954. *Trans. N. Y. Acad. Sci. Ser. II*, 17: 157—166.

Emmons, C.W. (ed.), 1960a. *Second Conference on Medical Mycology*, New York Academy of Science, *Ann. N. Y. Acad. Sci.*, 89: 1—282.

Emmons, C. W., 1960b. *Mycologia*, 52: 669—680.

Emmons, C. W., Binford, C. H. and Utz, J. P., 1970. *Medical Mycology*, Lea and Febiger, Philadelphia.

English, M. P., 1965. *Trans. Br. mycol. Soc.*, 48: 219—234.

English, M. P., 1967. *Sabouraudia*, 6: 218—227.

Gresham, G. A. and Burns, M., 1960. In R. Rook (ed.), *Progress in the Biological Sciences in Relation to Dermatology*, Cambridge University Press.

Gresham, G. A. and Whittle, C. H., 1961. *Sabouraudia*, 1: 30—33.

Howard, D. H., 1959. *J. Bact.*, 78: 69—78.

Howard, D. H., 1964. *J. Bact.*, 87: 33—38.

Howard, D. H., 1965. *J. Bact.*, 89: 518—523.

Howard, D. H., 1967. *J. Bact.*, 93: 438—444.

Howard, D. H. and Otto, V., 1969. *Sabouraudia*, 7: 186—194.

Jennison, R. F., 1966. In H. I. Winner and R. Hurley (eds.), *Symposium on Candida Infections*, p. 102—111, Livingstone, Edinburgh and London.

Johnson, P. T., 1968. *An Annotated Bibliography of Pathology in Invertebrates other than Insects*, Burgess, Minneapolis.

Kapica, L. and Blank, F., 1957. *Dermatologica (Basel)*, 115: 81—105.

Kunert, J., 1968. *Mykosen*, 11: 153—162.

Kunert, J., 1972. *Sabouraudia*, 10: 6—13.

Kwon-Chung, K. J., 1967. *Sabouraudia*, 6: 168—175.

Larsh, H. W. and Shepard, C. C., 1958. *J. Bact.*, 76: 557—563.

Louria, B. D. and Brayton, R. G., 1964. *Proc. Soc. exp. Biol. Med.*, 115: 93—98.

McGinnis, M. R. and Hilger, A. E., 1972. *Sabouraudia*, 10: 230—236.

Mackenzie, D. W. R., 1964. *Sabouraudia*, 3: 225—232.

Mackinnon, J. E. and Artagaveytia-Allende, R. C., 1948. *J. Bact.*, 56: 91—96.

Minocha, Y., Pasricha, J. S., Mohapatra, L. N. and Kandhari, K. C., 1972. *Sabouraudia*, 10: 79—82.

Montes, L. F. and Wilborn, W. H., 1968. *J. Bact.*, 96: 1349—1356.

Moss, E. M. and McQuown, A. L., 1960. *Atlas of Medical Mycology*, Williams and Williams, Baltimore.

Raubitschek, F., 1961. *Sabouraudia*, 1: 87—90.

Rifkind, D., Marchioro, T. L., Schnek, S. S. and Hill, R. B., 1967. *Am. J. Med.*, 43: 28—38.

Rippon, J. W. and Scherr, G. H., 1959. *Mycologia*, 51: 902—914.

Romano, A. H., 1966. In G. C. Ainsworth and A. S. Sussman (eds.), *The Fungi*, Vol. 2, Academic Press, London.

Scherr, G. H. and Weaver, R. H., 1953. *Bact. Rev.*, 17: 51—92.

Silverman, F. N., Schwarz, J. and Lahey, M. E., 1955. *Am. J. Med.*, 19: 410—459.

Sindermann, C. J., 1970. *Principal Diseases of Marine Fish and Shellfish*, Academic Press, New York.

Stuart, M. R. and Fulier, H. T., 1968. *Nature, Lond.*, 217: 90—92.

Taschdjian, C. L. and Kozinn, P. J., 1961. *Sabouraudia*, 1: 73—82.

Unestam, T., 1973. *Rev. Med. Vet. Myc.*, 8: 1—20.

Winner, H. I., 1966. In H. I. Winner and R. Hurley (eds.), *Symposium on Candida infections*, p. 6—11, Livingstone, Edinburgh and London.

Winner, H. I. and Hurley, R., 1964. *Candida albicans*, Churchill, London.

Yu, R. J., Harmon, S. R. and Blank, F., 1968. *J. Bact.*, 96: 1435—1436.

Yu, R. J., Harmon, S. R. and Blank, F., 1969. *J. invest. Derm.*, 53: 166—171.

Neutral Symbioses with Animals

Chapter 6

Neutral Fungi

Many fungi that are intimately associated with animals, particularly invertebrates, are neither obviously antagonistic nor mutualistic but are never found in a free-living form away from their hosts. Other fungi, that for the greater part of their life are free-living, frequently require a transient association with an animal, in the form of passage through its gut, in order to complete their life cycle. Such 'passage fungi' are commonly associated with warm-blooded animals. These two biological groups of fungi together form a large, heterogeneous body of obligately symbiotic neutral species that share the common characteristic of being dependent on an animal host while at the same time not affecting the metabolism of the latter to any degree. Although the neutral nature of most of these fungi is well established, some fall within this category only because they have not been examined critically for evidence of mutualism or antagonism.

There are, in addition, other fungi that can occupy ecological niches on or within the bodies of animals for indefinite periods also without producing any positive or negative effects on the host. The fungal flora of the skin, gut and mucous membranes of mammals contain species of this kind. These too might be regarded as neutrals, but for the majority there is no absolute requirement for the association and they can also be found free-living.

1. The Laboulbeniales

The Laboulbeniales (Laboulbeniomycetes) is a little-known order in the Ascomycotina comprising fungi that are ectosymbiotic upon insects, mostly Coleoptera, and also on myriopods and mites. Over 1,500 species have been described, yet there is little information available as to details of their nutritional relationships with their hosts (Shanor, 1955; Madelin, 1963).

Infection is brought about by ascospores which germinate on the integument of the host and give rise to a thallus anchored to the surface of the host by means of a foot (Figure 34). The thallus is of limited growth and its final size varies with the species, so that it may either be minute and composed of only a few cells, or be massive and made up of an indefinite number of cells. It may bear appendages and male and female reproductive organs. Species can be monoecious, dioecious, or have female organs only, the latter then developing parthenogenetically. Ascospores are produced within a perithecium and their dissemination is effected by host to host contact, although in some cases infection occurs through contact with ascospore-contaminated soil (Lindroth, 1948). Only one species has so far been grown in axenic culture, *Stigmatomyces ceratophorus* from the lesser house fly *Fannia canicularis* having been grown to a relatively large size (20 cells) on detached fly wings placed on nutrient agar, and the major questions as to what their nutrient requirements are and how they are satisfied remain unanswered (Whisler, 1968*a*).

Considerations of their possible modes of nutrition usually involve speculation on the physical nature and possible function of the basal foot. The entire thallus is usually enclosed by a continuous, thin, supposedly impermeable membrane and the cell walls interior to this are often very thick (Faull, 1912). These structures presumably prevent desiccation but possibly also act as an effective barrier to nutrient movement into the thallus through its surface (Thaxter, 1914; Benjamin and Shanor, 1950). If this is so, then the only alternative route for the entry of water and nutrients is through the membranes of the foot cells.

In some species penetrant organs arise from the foot, pass through the integument and reach the soft tissues (Chatton and Picard, 1909; Whisler, 1968*b*; Richards and Smith, 1956). For example, in *Herpomyces stylopygae* slender hyphae pass to the hypodermis where they swell to form bulbous organs (Figure 32). Growth rate studies show that this species probably obtains the bulk of its nutrients through these structures and that growth of thalli ceases when the cuticle is explanted. The limiting membrane of the hypodermis may be stretched by the growth of these assimilative organs but it is not ruptured and the fungus seems to be incapable of breaking into the haemocoel. Other species can effect deeper penetration and *Trenomyces histophthorus* produces bulbs and nodular tubes that enter the host's fat body (Figure 33). Neither of these species, nor *Stigmatomyces ceratophorus* which also penetrates into tissues underlying the integument, seem to have any measurable effect on host growth.

The great majority of species do not, however, make any obvious penetration of even the outer layers of the cuticle, and the foot seems to serve merely as a holdfast. There are no reports of physical damage to the cuticle during thallus growth and there is no correlation between thallus size and the relative size of the foot, even massive thalli having only a very small foot. In addition, large foot cells are not found in species which grow on spines or hairs, where the availability of nutrients from the host might be expected to be low. The role of the

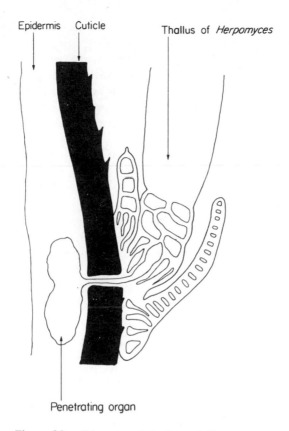

Epidermis   Cuticle   Thallus of *Herpomyces*

Penetrating organ

Figure 32   Diagram of thallus of *Herpomyces*
stylopygae on the antenna of the cockroach
Blatta orientalis, showing inflated penetrating
organ within the epidermis. Redrawn from
Richards and Smith, 1956

non-penetrant foot as a provider of nutrients must, therefore, be
critically questioned but there are a number of ways in which it could,
possibly, carry out this function. First, species with simple holdfasts
might also possess minute, submicroscopic penetrating structures, but
no electron microscope studies have so far been made on the
Laboulbeniales to confirm or eliminate this possibility. Second, thalli
may be exploiting cuticular secretions, containing waxy or fatty
materials, as nutrient sources with the foot cells acting as absorption
points. Provided that the secretions were rich enough in nutrients then
a small holdfast could support growth of a large thallus. Third, the
ground substance of the cuticle might be degraded, but very slowly, by
extracellular enzyme production by the foot cells. Gradual removal of
materials during a long period, and over a relatively wide area, would

lead to the formation of microscopic lacunae within the cuticle rather than to severe localized erosion. Such activity could also induce a continuous flow of materials, in the form of cuticle components, towards the sites of removal. It is also conceivable that, in at least some species, the thallus membrane and cell wall do not present a totally impermeable barrier. Thalli growing on aquatic insects or on hosts from damp habitats could then obtain dissolved nutrients from the environment through their surfaces.

One of the many curious features of some of the Laboulbeniales is their high degree of specificity (Shanor, 1955). Many species occur on a range of hosts, for example on both beetles and mites, but other show extreme restriction. They may be limited to a single host species, occur on one or other sex of a particular host, or be confined to particular positions on their hosts (Table 5; Figure 34). The bases for species and sex specificity are not clear but might obviously be due to the physical,

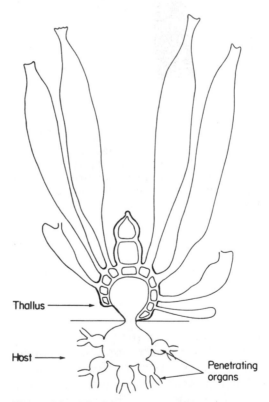

Figure 33   Thallus of *Trenomyces histo-phthorus* on the body of a chicken flea, showing penetrating organs; Redrawn from Chatton and Picard, 1909

Table 5

Distribution of *Laboulbenia* species on males and females of the beetle *Bembidion picipes* (after Benjamin and Shanor, 1952)

Species of fungus	L. vulgaris		L. truncata		L. perpendicularis		L. bembidio-palpi		L. odobena		L. tapirina	
Sex of insect	♀	♂	♀	♂	♀	♂	♀	♂	♀	♂	♀	♂
Number of insects with thalli	87	122	0	36	24	0	0	27	37?	58	50	50

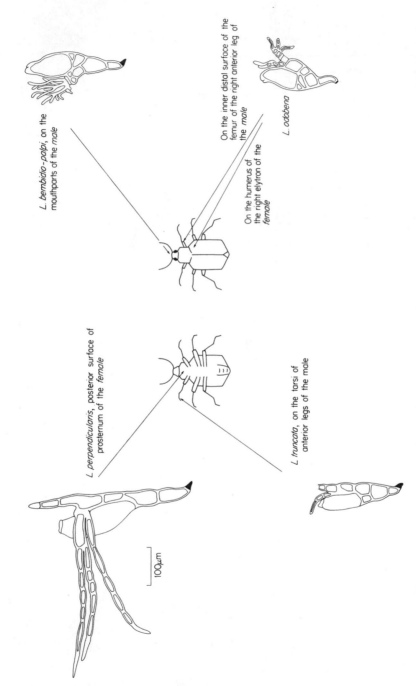

L. bembidio-palpi, on the mouthparts of the *male*

On the inner distal surface of the femur of the right anterior leg of the *male*

On the humerus of the right elytron of the *female*

L. odobena

L. perpendicularis, posterior surface of prosternum of the *female*

L. truncata, on the tarsi of anterior legs of the male

100 μm

Figure 34 Position and sex specificity of some *Laboulbenia* species on the beetle *Bembidion picipes*. Composite diagram based on drawings and photographs in Benjamin and Shanor, 1952; by permission of the *American Journal of Botany*

nutritional or hormonal characteristics of the animal. Some examples of positional specificity have been shown to be due to behavioural features, such as grooming or mating, that lead to a particular distribution of spores on the host's body (Benjamin and Shanor, 1952; Shanor, 1952). Other instances cannot be explained in these terms and spores evenly distributed over the body sometimes will only germinate on particular parts of it (Richards and Smith, 1955). In *Chitinomyces*, a genus which infects water beetles, instances occur where the exact position of the thalli of a particular species on its host is invariable, and corresponds exactly in different host individuals, even when these occur in widely separated habitats. This positional specificity is such that the identity of the fungus can be predicted if the host and the position of the fungus upon it are known (Hincks, 1960).

2. The Trichomycetes

The Trichomycetes are a group of fungi which, with the exception of a single genus, are symbiotic within the digestive tracts of Arthropods. Although a number of species have been grown in axenic culture they have never been found free-living in nature (Lichtwardt, 1960). Their simplicity of form, together with their mode of sexual reproduction, indicates that some of them have a strong affinity with, or indeed are, members of the Zygomycotina (Lichtwardt, 1973).

All species have septate or aseptate, cylindrical or filamentous, branched or unbranched thalli attached to the host by a basal holdfast (Figures 35 and 36). The group is made up of four orders, the Harpellales, Asellariales, Eccrinales and Amoebidiales but the taxonomic relationships between them are obscure. For example, the Harpellales have chitinous cell walls while those of the Eccrinales are composed principally of cellulose. The Amoebidiales have neither chitin nor cellulose in their walls but these contain instead proteins, glucosamine and galactose (Trotter and Whisler, 1965; Lichtwardt, 1973). Immunological studies suggest that the Harpellales resemble the Kickxellaceae in the Mucorales, but that the Amoebidiales have no relationship to either the Harpellales or to the Kickxellaceae. It has been intimated that the Amoebidiales may be protozoa not fungi (Whisler, 1962, 1965; Sangar, Lichtwardt, Kirsch and Lester, 1972).

Trichomycetes are found in the midgut, hindgut and rectum of their hosts, the only exception being *Amoebidium*, the thalli of which are external on aquatic crustacea or insects. Infection of the gut is brought about by the ingestion of asexual or sexual spores that subsequently germinate, become anchored to the intestinal wall, and produce a thallus. When the host moults, the gut lining, together with any attached thalli, is shed. Reinfection takes place if the host eats its moult or through thalli being borne on gut nematodes which remain inside the

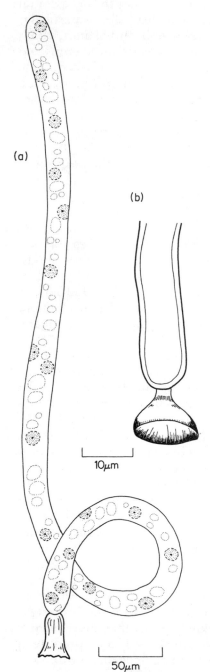

(a)

(b)

10μm

50μm

Figure 35 Trichomycetes: (a) thallus of *Enterobryus elegans* from gut of a millepede; (b) bell-shaped holdfast of *E. borariae*. Redrawn from Lichtwardt, 1954, 1958; by permission of *Mycologia*

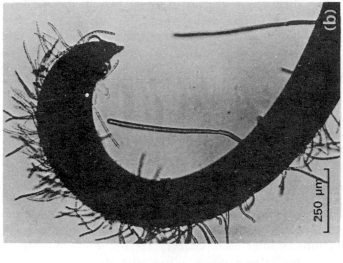

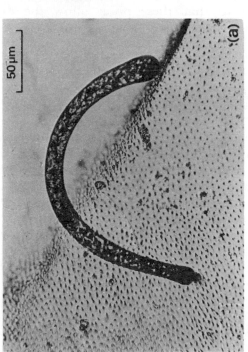

Figure 36 *Enterobryus elegans* from the millipede *Spirobolus americanus*: (a) young thallus attached to the chitinous lining of the hindgut; (b) part of a gut nematode bearing numerous thalli. From Lichtwardt, 1954; by permission of *Mycologia*

host during the moult (Figure 36) (Lichtwardt, 1954; Whisler, 1963). In larvae of Diptera, thalli may not be shed at moulting, at least from the midgut. This is lined by the peritrophic membrane which is derived from a band of cells around the stomodaeal valve. The membrane continually grows throughout the larval moults and is not shed, but lies free within the midgut lumen and projects slightly into the hindgut. Thalli attached to this membrane complete their development before its growth carries them into the hindgut where the membrane is continually digested. Young thalli are, therefore, distributed in the anterior region of the midgut and mature thalli in its posterior region (Moss, 1970; Reichle and Lichtwardt, 1972).

In contrast to the situation in the Laboulbeniales there is no speculation as to the function of the holdfast, which is simply an anchor for the thallus. Trichomycete holdfasts may be either cellular or non-cellular. Non-cellular holdfasts have no cytoplasmic core so that there is no bridge for nutrient movement between host and fungus, nor is there any lysis of the gut lining beneath the point of attachment (Lichtwardt, 1954). The wall of the thallus immediately adjacent to the holdfast may, however, contain numerous fine perforations (Whisler and Fuller, 1968; Manier and Grizel, 1972). Electron microscope studies indicate that these pores allow the extrusion of adhesive holdfast material and that they do not have an absorptive function (Figure 37b). Cellular holdfasts have a wide range of structure and can be simple, digitate, or branched, and may possess or lack a surrounding mucilaginous matrix (Figure 37a). Despite their often complex structure they nevertheless seem not to be absorbing organs, even in situations when they penetrate the peritrophic membrane (Whisler, 1963; Moss 1970; Lichtwardt, 1973).

Trichomycetes obtain all their nutrients from the fluids of the digestive tract or, in the case of *Amoebidium*, from the water in which the host lives. Attachment to the gut is not essential and thalli can grow epiphytically on the thalli of other Trichomycetes or on the cuticle of gut-inhabiting nematodes. The exact nature of their nutrient requirements and of those factors which might cause them to be confined to their hosts have not yet been determined. Only one endosymbiotic genus, *Smittium*, has been successfully grown in axenic culture (Clerk, Kellen and Lindegren, 1963; Lichtwardt, 1964; Farr and Lichtwardt, 1967; Williams and Lichtwardt, 1972). *Smittium* species are members of the Harpellales and some can be grown on protein-rich, undefined media. The nutrient requirements of one fungus, *S. culisetae*, have been partially determined and it develops well on media containing tryptone, glucose and thiamine. It utilizes a wide range of hexoses and disaccharides, as well as glycerol, as major carbon sources and can hydrolyse starch. Inorganic nitrogen sources, asparagine, methionine and casamino acids are all inferior to tryptone in satisfying its nitrogen

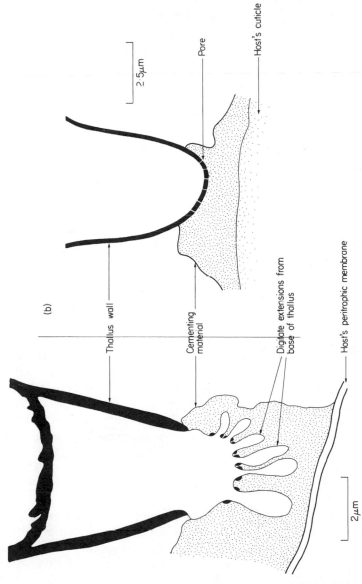

Figure 37 Trichomycete holdfasts: (a) *Harpella melusinae* from midgut of a black-fly larva; (b) *Amoebidium parasiticum* on the cuticle of *Daphnia*. (a) Based on an electronmicrograph in Reichle and Lichtwardt, 1972; by permission of *Archiv für Mikrobiologie*. (b) Based on electronmicrographs in Whisler and Fuller, 1968; by permission of *Mycologia*

Pore

Host's cuticle

2·5μm

Thallus wall

Cementing material

Digitate extensions from base of thallus

Host's peritrophic membrane

2μm

(a)

(b)

requirements. If reconstituted tryptone, made up from known compounds, is used then growth in media containing it is poor. Commercial
tryptone therefore appears to contain non-vitamin factors which
promote growth. The ectosymbiont *Amoebidium parasiticum* has also
been cultured axenically and good growth is obtained on a tryptone—
glucose—thiamine medium (Whisler, 1960, 1962). Fructose and mann-
ose can successfully replace glucose as major carbon sources but
disaccharides and polysaccharides are not utilized.

A. parasiticum has a complicated pattern of development which is
integrated with that of its host (Whisler, 1960, 1962, 1965, 1968*b*). Its
cylindrical thallus becomes transformed to a sporangium on maturation
and either non-motile spores or amoeboid cells are released. It is the
amoeboid cells that give rise to the resistant phase of the fungus. After
a few hours of movement they encyst and, later, non-motile spores are
formed within each cyst. Formation of amoeboid cells only occurs
when the host moults or dies (Tuzet and Manier, 1951). In axenic
culture thalli normally produce only non-motile spores, but a water-
soluble, thermostable amoebagenic factor has been obtained from
Daphnia which induces amoeboid cell formation in axenic culture. Such
a factor might be released naturally at moulting or death and would
then regulate the pattern of fungal development. Similar effects can be
induced by high calcium levels but not by high concentrations of other
divalent cations. The presence of calcium may in some way interfere
with normal cell wall formation, so that it is not possible to produce
non-motile spores but only naked protoplasts.

Another member of the Amoebidiales, *Paramoebidium*, closely
resembles *Amoebidium* in having an amoeboid cell-cyst stage but this
species is endosymbiotic within the hindgut. Moults of infected insect
larvae are covered with large numbers of amoeboid cells which may
then encyst and be ingested by a new host. Axenic cultures of
Paramoebidium can be used as inoculum to produce ectosymbiotic
infections, and on these grounds it has been suggested that this genus is
synonymous with *Amoebidium* (Whisler, 1965).

3. Passage Fungi

When fungi require passage through an animal then, during passage,
physical or physiological changes occur in the fungus that are necessary
for normal phenotypic development to subsequently take place.
Sometimes passage is an absolute requirement but frequently, although
development will take place without it, passage enhances development.

The spores of many coprophilous fungi will not germinate or will
germinate only very slowly on nutrient agar, but will do so readily after
passage through an animal. Passage breaks the natural dormancy of the
spores and allows them to germinate and give rise to a mycelium. When

they pass through the gut, particularly of mammals, they are subjected to a relatively high ambient temperature, to the action of enzymes, and to wide fluctuations in pH. Any one of these factors or a combination of all of them may be responsible for dormancy-breaking. The temperature of the animal alone may be an important factor. For example, ascospores of *Ascobolus* species can be activiated artificially by subjecting them to short incubation periods at mammalian blood temperature. Treatment with dilute alkalis or alkaline pancreatin will similarly activate spores of other coprophilous fungi (Yu, 1954; Webster, 1970). It has been suggested that, since large numbers of coprophilous species occur in the dung of birds and mammals but relatively few in that of reptiles and amphibia, the warm-blooded condition has been of importance in the evolution of the coprophilous habit. Although activation of spores is well established, little is known of the restraints on germination that are removed during passage. Natural development subsequent to activation usually takes place after the faeces have been voided, although spores of some species germinate within the rectum.

Basidiobolus ranarum is one of the few species that are found consistently in the dung of cold-blooded animals, principally frogs (Levisohn, 1972; Drechsler, 1956). Its conidia may be ingested by insects or adhere to their body surface, but in either case no further development takes place until the insect is eaten by a frog. The conidia then divide within the frog's gut to form spherical cells which germinate and produce mycelium within the voided faeces. Closely related species can cause mycetomas in mammals, including man, and these fungi, unlike *B. ranarum*, grow readily at mammalian blood temperature (Greer and Friedman, 1966).

There are a number of other fungi which, although not coprophiles, require their spores to pass through an animal for germination to take place (Sussman and Halvorsen, 1966). Examples include some Hymenomycetes, the Ascomycotina species *Onygena corvina*, and some Gasteromycetes. These require passage through slugs, owls, and rodents respectively. The basidiospores of some species in the Phallales may require passage through insects.

Bibliography

Benjamin, R. K. and Shanor, L., 1950. *Am. J. Bot.*, 37: 471—476.
Benjamin, R. K. and Shanor, L., 1952. *Am. J. Bot.*, 39: 125—131.
Chatton, E. and Picard, F., 1909. *Bull. Soc. Mycol. France*, 25: 147—170.
Clerk, T. B., Kellen, W. R. and Lindegren, J. E., 1963. *Nature, Lond.*, 197: 208—209.
Drechsler, C., 1956. *Mycologia*, 48: 655—676.
Farr, D. F. and Lichtwardt, R. W., 1967. *Mycologia*, 59: 172—182.
Faull, J. H., 1912. *Ann. Bot.*, 26: 325—355.

Greer, D. L. and Friedman, L., 1966. *Sabouraudia,* 4: 231–241.

Hincks, W. D., 1960. *Naturalist, Hull,* 874: 97–102.

Levisohn, I., 1927. *Jb. wiss. Bot.,* 66: 513–555.

Lichtwardt, R. W., 1954. *Mycologia,* 46: 564–585.

Lichtwardt, R. W., 1958 *Mycologia,* 50: 550–561.

Lichtwardt, R. W., 1960. *Mycologia,* 52: 410–428.

Lichtwardt, R. W., 1964. *Am. J. Bot.,* 51: 836–842.

Lichtwardt, R. W., 1973. *Mycologia,* 65: 1–20.

Lindroth, C. H., 1948. *Svensk. bot. Tidskr.,* 42: 34–41.

Madelin, M. F., 1963. In E. A. Steinhaus (ed.), *Insect Pathology* Vol. 2, p. 233 – 271, Academic Press, New York.

Manier, J. F. and Grizel, H., 1972. *C.r. hebd. Scéanc. Acad. Sci. Paris,* 274: 1159–1160.

Moss, S. T., 1970. *Trans. Br. mycol. Soc.,* 54: 1–13.

Reichle, R. E. and Lichtwardt, R. W., 1972. *Arch. Mikrobiol.,* 81: 103–125.

Richards, A. G. and Smith, M. N., 1955 *Biol. Bull.,* 108: 206–218.

Richards, A. G. and Smith, M. N., 1956. *Ann. ent. Soc. Am.,* 49: 85–93.

Sangar, V. K., Lichtwardt, R. W., Kirsch, J. A. W. and Lester, R. N., 1972. *Mycologia,* 64: 342–358.

Shanor, L., 1952. *Am. J. Bot.,* 39: 498–504.

Shanor, L., 1955. *Mycologia,* 47: 1–12.

Sussman, A. S. and Halvorsen, H. O., 1966. *Spores, their Dormancy and Germination,* Harper and Row, New York and London.

Thaxter, R. 1914. *Bot. Gaz.,* 58: 235–253.

Trotter, M. J. and Whisler, H. C. 1965. *Can. J. Bot.,* 43: 869–876.

Tuzet, O. and Manier, J. F. 1951. *Annls. Sci. nat. (Zool.) Sér. 11.* 13: 351–364.

Webster, J. 1970. *Trans. Br. mycol. Soc.,* 54: 161–180.

Whisler, H. C. 1960. *Nature, Lond.,* 186: 732–733.

Whisler, H. C. 1962. *Am. J. Bot.,* 49: 193–199.

Whisler, H. C. 1963. *Can. J. Bot.,* 41: 887–900.

Whisler, H. C. 1965. *J. Protozool.,* 13: 183–188.

Whisler, H. C. 1968a. *Mycologia,* 60: 65–75.

Whisler, H. C. 1968b. *Devl. Biol.,* 17: 562–570.

Whisler, H. C. and Fuller, M. S. 1968. *Mycologia,* 60: 1068–1079.

Williams, M. C. and Lichtwardt, R. W. 1972. *Mycologia,* 64: 806–815.

Yu, C. C. C., 1954. *Am. J. Bot.,* 41: 21–30.

Mutualistic Symbioses with Animals

Chapter 7

Mutualistic Symbioses with Insects

Many insect species have Hyphomycetes, Ascomycotina, or Basidiomy-cotina constantly associated with them. Usually the insects depend upon higher plant materials ultimately for their nutrients but, lacking the ability to break these down and digest them themselves, allow fungi to do so and then feed on the fungal mycelium. When the insects move from one habitat to another the fungi are taken with them, and efficient mechanisms have been evolved by both partners in the symbiosis to ensure successful transmission of the fungi. Although some other scattered examples are known, studies on these kinds of symbioses have been concentrated upon four major ecological groups of insects. These are wood-inhabiting beetles and wasps, leaf-cutting ants, scale insects, and insects with symbiotic fungi that live entirely within their bodies (Francke-Grosmann, 1967; Graham, 1967; Hartzell, 1967).

1. Fungi of Wood-Inhabiting Insects

Wood, whether living, moribund, or dead, provides a favourable habitat for many insects which not only bore into it to create an abode but also utilize it as a food source. It has been pointed out that woody tissue is a particularly unpromising food source for insects in that its main components, lignin and cellulose, are resistant to degradation by insect enzymes. In addition, woody substrates are deficient in sterols and vitamins, compounds which insects cannot synthesize for themselves (Baker, 1963). The success of many wood-inhabiting insects is due to the evolution of symbioses between them and fungi which enable all their nutrient requirements to be satisfied. The symbiotic fungi are sometimes housed, and grow inside, specialized host organs but they always remain saprotrophic and extracellular, and are never biotrophic within living host cells. These symbioses are well documented but certain aspects of their biology are still imperfectly understood, due mainly to a paucity of experimental investigations carried out under controlled conditions.

Ambrosia and Blue-Stain Fungi

The xylophagous ancestors of present day wood-inhabiting beetles possibly entered wood only after its decay had begun, relying on the activity of saprotrophic micro-organisms to reduce the wood to a palatable and utilizable condition (Schedl, 1958). Later, there presumably evolved xylophagous beetles which could utilize dead, but undecayed, wood by virtue of endosymbiotic, mutualistic microorganisms. A further advance then took place when phloeophagous forms evolved which were able to colonize living phloem, cambium, or outer sapwood, A final specialization occurred when beetles re-entered deep wood tissue and evolved into the present wood-inhabiting Coleoptera, also known as ambrosia beetles or xylomycetophagous beetles. These are wood-borers but not wood-feeders, symbiotic fungi which grow on their tunnel walls acting as their primary food source. Phloeophagy, xylomycetophagy and, to a lesser extent, xylophagy are all found within the two most successful and widespread groups of wood-inhabiting beetles, the Scolytidae and the Platypodidae.

Xylomycetophagous beetles usually inhabit low-vitality wood, for example recently fallen timber. Flying adults alight upon this and boring begins. In the Scolytidae usually, but not always, it is the female that initiates tunnelling, while in the Platypodidae it is the male (Francke-Grosmann, 1967). After a short period of tunnelling, copulation takes place outside the tunnel and boring is then resumed, invariably by the female. Breeding tunnels are excavated and eggs are laid in them.

The tunnels are kept free from debris and faeces and become lined with a fungal layer upon which the beetles and their larvae constantly browse. Wood fragments are always present in the gut of feeding adults, and usually also in the larvae, but whether this is an important factor in their diet is uncertain, although its starch content is depleted during its passage through the gut. After pupation young adults feed for a time on the fungal layer, then cease feeding and void their digestive tracts. They pass out of the tunnels, fly to a fresh site, excavate a new tunnel and mate. They do not feed until a fungal layer is established within the new tunnels. During this period of absence of fungal food they derive nourishment from autolysis of their flight muscles.

Within the tunnels the ambrosia fungi form a thin, continuous palisade over the walls, or grow as separate or confluent sporodochia. A single species or a mixture of several species may be present and whether one or more fungi take part in the symbiosis seems to depend on the species of insect involved. Older tunnels become invaded by alien fungi which compete with, and eventually eliminate, the symbiotic species. The palisades or sporodochia consist of erect hyphae which frequently bear monilioid chains of conidia or terminal chlamydospores. Beneath these hyphae the mycelium penetrates the tunnel walls to a depth of a few millimetres with occasionally deeper penetration occurring. Wood fragments are distributed within the fungal layer.

The taxonomy of ambrosia fungi is somewhat confused and, although most may be placed within four hyphomycetous genera, *Ambrosiella, Raffaelea, Monacrosporium* and *Phialophoropsis*, it is clear that many more genera are involved including *Fusarium, Cephalosporium, Candida* and *Graphium*. Yeasts such as *Ascoidea* and *Endomycopsis* may also be of some importance (Batra and Francke-Grosmann, 1961; Batra, 1963; Batra and Francke-Grosmann, 1964; Batra, 1967; Baker and Norris, 1968). Contributing to this confusion is the fact that among the primary ambrosia fungi may be found secondary or auxiliary ambrosia fungi which, although not perhaps acting as a major nutrient source for the beetles, supplement the effects of the primary fungi. Auxiliary species often appear in the galleries during pupation and may, perhaps, only be eaten by newly emerged adults. Some fungi that cause blue stain of wood, for example *Ceratocystis, Botryodiplodia* and *Leptographium,* may act as primary or auxiliary ambrosia fungi but this is open to doubt (Francke-Grosman, 1967).

Ambrosia fungi can be grown in axenic culture but most, unlike their non-ambrosial counterparts, are extremely pleomorphic and either assume a yeast form or grow as a rapidly spreading mycelium rather than displaying their natural ambrosial growth habit (Batra, 1967). Ambrosia-type growth has occasionally been observed in axenic culture but this occurs infrequently and those factors which induce it are not understood. They may be nutritional, although injury to mycelia and the presence of adult beetles or larvae can, apparently, stimulate sporodochium formation. The nutritional requirements and physiological properties of ambrosia fungi have not been rigorously studied despite the ease with which they can be cultured, and the little information that is available serves only to confuse. Species seem to differ widely from one another in their physiology and, apart from an obvious ability to break down one or more components of wood, seem to have few physiological characteristics in common. This is true of even taxonomically closely related fungi. For example, in axenic culture *Endomycopsis platypodus* will utilize nitrate as sole source of nitrogen while the congeneric species *E. fasciculata* will not (Batra, 1963; Baker and Kreger-van Rij, 1964). Some species are heterotrophic for vitamins while others are autotrophic, some will utilize uric acid as a nitrogen source and others will not. *Monilia ferruginea*, an ambrosia fungus of Scolytids, requires casein, peptone or olive oil for sucessful spore germination on agar media, and will produce sporodochia only in the presence of these complex nutrients (Baker, 1963).

Some confusion also exists concerning the question of the possible specificity of fungus—beetle associations. It has been categorically stated that each beetle species is associated with a particular ambrosia fungus, or particular complex of fungi, which is essential for normal development of the brood. It has also been suggested that closely related beetle species are associated with closely related fungi, or different strains of the same fungus, and that these fungi are not only highly adapted to their insect partner but also to the environment in which they are normally found (Francke-Grosmann, 1967). Although this may well prove to be true for a number of associations, available evidence points to this being by no means always the case (Buchanan, 1941). Broods of the Scolytid *Xylosandrus germanus* have been sucessfully reared to adulthood on the non-ambrosial fungi *Ceratocystis ulmi, C. piceae,* and a *Pestalotia* species. At least some beetles are therefore not rigidly adapted to utilize one particular fungus as a food source.

Ambrosial fungi are transported from place to place by adult beetles and the latter have anatomical modifications that allow them to carry and maintain the symbionts in a viable condition. In beetles of the Scolytidae the fungi are contained in specialized integumental organs, the mycetangia, which are basically depressions or invaginations of the body surface (Figure 38). Only the primary and not the auxiliary fungi are found in them (Batra, 1967). Mycetangial fungi are present as spores or as yeast-like cells and are retained within the organs by means of hairs or spines that project from the host's integument. The location of mycetangia on the body and their range of structure vary but are constant for any beetle species (Table 6). Generally, mycetangia are present in only one sex of the insect, that which bores into wood first. All contain an oily secretion and, where mycetangia are sac-like, they may have heavy musculature (Lowe, Giese and McManus, 1967). In the beetle *Dendroctonus frontalis*, mycetangial secretions are produced by

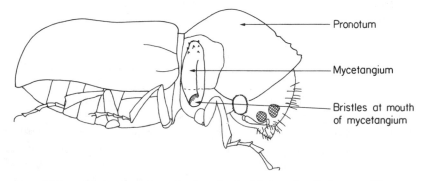

Pronotum

Mycetangium

Bristles at mouth of mycetangium

Figure 38   Adult female of the Scolytid beetle *Xyloterus lineatus* showing position of one of the paired mycetangia. Redrawn from Franke-Grosmann, 1967; copyright Academic Press Inc

Table 6
Nature and location of mycetangia in some wood-inhabiting beetles
(after Francke—Grosmann, 1967)

Type of mycetangium	Beetle species
	Scolytidae
Paired glandular tubes in prothorax	*Xyloterus lineatus*
Enlargement of precoxal cavity	*Pterocyclon brasiliense*
Intersegmental pouches between pro- and mesonotum	*Xylosandrus germanus*
Membraneous pouches at base of mandible	*Xyleborus fornicatus*
Pharyngeal pouches	*Premnobius cavipennis*
Pouches at base of elytra	*Xyleborus saxeseni*
	Platypodidae
A pair of small hollows on pronotum	*Platypus hintzi*
Numerous small punctures on pronotum	*P. wilsoni*

two kinds of glandular cell, and it has been suggested that these secretions serve as regulators of the growth of ambrosial fungi within the mycetangia and also act as inhibitors of the growth of non-ambrosial fungi (Happ, Happ and Barras, 1971). During boring, secretion increases so that the oily liquid oozes out of the mycetangia, carrying fungal cells with it. This process is aided by contraction of the muscles and movement of the mycetangial hairs. The fungi then begin to grow in the tunnels, perhaps initially utilizing the secretions smeared on the walls. Originally the oil-filled cavities of the body surface may have had the primary function of providing a medium to lubricate the insect's body during boring, and this would also have acted as a water repellent against excessive sap flow occurring in the wood as a response to wounding. During evolution of the symbiosis, fungi were, perhaps, selected which could either utilize or tolerate these secretions, and so eventually came to grow within the secreting organs. How ambrosia fungi enter the mycetangia from the tunnel wall after pupation has taken place is not known.

In the Platypodidae mycetangia, if present, have a very simple form, being small pits or notches in the integument (Figure 39). Usually the male initiates tunnelling, but where true mycetangia have been found these are present only on the female (Francke-Grosmann, 1967). Fungi of Platypodids are also commonly associated with oily secretions either from glandular hairs or from tubules opening to the body surface. In such instances both sexes carry the fungi, but the secretions may be more evident on the females.

The broad bases for the fungus–beetle symbiosis are reasonably clear. The ambrosia fungus is inoculated directly into a favourable

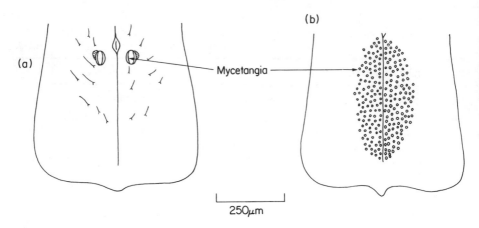

Figure 39   Position of mycetangia in female Platypodid beetles. (a) *Platypus hintzi*, posterior end of pronotum, two mycetangia; (b) *P. oxyurus*, posterior end of pronotum, numerous mycetangia. Redrawn from Francke-Grosmann, 1967; copyright Academic Press Inc

habitat in wood and is also protected and propagated within the mycetangia, while the beetle derives a major part of its nutritional requirements from eating the fungus. However, exact details of the symbiosis, particularly those of fungal and beetle nutrition, are not fully understood. The beetles do not culture the fungi 'by intent', but it is likely that ambrosia fungi cannot colonize wood in the absence of their insect partners and that some will not grow in the absence of adult beetles. The simplest possible picture of nutritional behaviour of the fungi in nature is one where they require lipid from insect secretions together with a suitable nitrogen source. Wood fragments impregnated with uric acid and voided by the beetles could satisfy the latter requirement in at least some fungi. Those fungi that are heterotrophic for vitamins may be supplied with these by the auxiliary fungi growing within the ambrosial mycelium.

Effects of the fungi on their insect partners have been most studied using the Scolytid beetle, *Xyleborus ferrugineus*. Adults have been reared from aseptic eggs on complex agar media usually containing, among other materials, a mixture of sucrose, protein and yeast extract, together with plant extracts rich in lipids, polysaccharides and vitamins (Saunders and Knoke, 1967). Females tunnel and feed within the medium but eggs produced by these asymbiotic adults are not viable. Oviposition of viable eggs only takes place after a mutualistic fungus, *Fusarium solani*, which is present in the oral mycetangia of symbiotic adults, is inoculated into the medium (Norris and Baker, 1967). This ambrosial fungus is therefore essential for reproduction, but not for growth, of *X. ferrugineus*. Its efficiency under some experimental conditions seems to depend on its capacity to produce an essential sterol (Chu, Norris and Kok, 1970; Kok, Norris and Chu, 1970). It has been found that fungus-free females supplied with cholesterol produce second generation larvae but that these will not pupate. However, *F. solani* produces ergosterol and this sterol allows growth, development and reproduction to proceed normally in asymbiotic beetles. In addition to providing nutrients in the broad sense ambrosia fungi may therefore also supply compounds that control normal development. A *Cephalosporium* and a *Graphium* species have been isolated from the mycetangia of *X. ferrugineus* and these too, when fed to adults or larvae, allow development to take place but are not as beneficial as *F. solani* (Baker and Norris, 1968).

As well as being associated with Scolytid and Platypodid beetles, ambrosia fungi have also been found in the larval galleries of some species of the Lymexylidae. Adult Lymexylids live free and deposit their eggs in bark crevices. The larvae bore into the wood and the ambrosia fungi grow in the tunnels which are usually moist with fermenting sap. In adults the symbiont is stored in paired mycetangia near the distal end of the ovipositor, so that when eggs are laid they become contaminated by the fungus. The hatched larvae then carry the propagules on their body surfaces. All Lymexylid fungi so far studied are species of the hemiascomycetous genus *Ascoidea* (Batra and Francke-Grosmann, 1961, 1964).

In contrast to xylomycetophagous beetles, phloeophagous beetles inhabit niches rich in easily assimilated food, and yet many of them are constantly associated with yeasts (Shifrine and Phaff, 1956). Certain blue-stain fungi also occur in their galleries. *Ceratocystis* and *Trichosporium* are the most common of these fungi which discolour wood but do not cause it to decay. The association between blue-stain fungi and phloeophagous beetles is probably not a symbiotic one in most cases, the fungi being merely accidentally transmitted by the insects as they move from one tree to another. However, in some instances there does seem to be a symbiosis and larvae consume the fungi during some stage of their development (Leach, Hodson, Chilton and Christensen, 1940). Where there is symbiosis the blue-stain fungi are transmitted in a yeast-like form in mycetangia that consist of either simple punctures in the integument, or of well-developed pouches or tubes near the mandibles or on the prothorax. As with true ambrosia fungi the fungal cells are suspended in an oily secretion. Larvae of phloeophagous species have been successfully reared aseptically, which implies that, for their part, the symbiosis is facultative rather than obligate. It may be significant with respect to this that no major physiological differences have been found between *Ceratocystis* species that are not normally associated with insects and those that are (Mathieson-Käärik, 1960).

Wood Wasp Fungi

Larvae of the Siricidae and Xyphidriidae inhabit weakened trees or freshly cut timber into which the adult female wasp inserts her eggs several centimetres below the surface of the wood by means of a long ovipositor. After hatching, the larvae tunnel more deeply into the wood and develop to the pupal stage in debris-packed tunnels. Adult females have at the base of the ovipositor, and opening into it, a pair of intersegmental pouches filled with the oidia or the short, branched, hyphae of a fungus. The fungal cells are suspended in mucus secreted by a pair of glandular organs at the sting base, and are retained within the pouches by means of inward-pointing bristles located near the exit of these organs (Figure 40a, b). During oviposition, contraction of the pouches squeezes out some of the fungal cells which adhere to the egg's

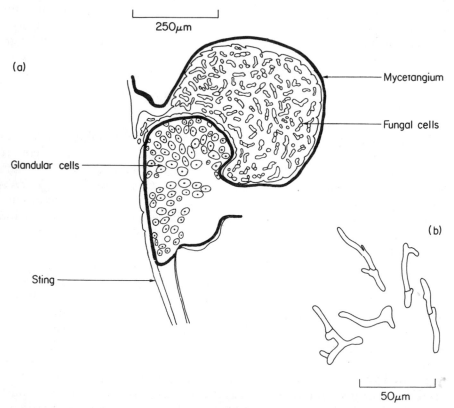

Figure 40   Wood wasp fungi: (a) vertical section through the base of the ovipositor of *Sirex juvencus*; (b) oidia from the mycetangium of *S. gigas*. (a) Redrawn from Francke-Grosmann, 1967; copyright Academic Press Inc. (b) Redrawn from Buchner, 1965; by permission of John Wiley & Sons

surface and pass out with it. All female wood wasps have these pouch-like mycetangia except in *Xeris spectrum* where they are apparently absent.

After deposition of the eggs the fungal cells rapidly give rise to a mycelium which permeates the wood, rapidly breaking down cellulose and lignin (Francke-Grosmann, 1967). When larvae hatch they therefore move about and feed in tissues that have been radically altered by fungal activity. Exactly how the larvae benefit from the presence or activity of the fungus is not clear. They may of course be xylomycetophagous, and there are reports of *Sirex* larvae being able to live for short periods on cultures of their associated fungus. It has also been shown that larvae possess enzymes which would allow them to digest mycelium. However, within wood the mycelium is sparse and it is difficult to visualize it as being capable of acting as a major source of nutrients for the larvae. Another possibility is that the larvae are primarily xylophagous but are partially or totally deficient in those enzymes necessary for the digestion of wood. Wood digestion by larvae would be made easier through the activity of fungi within it prior to its ingestion. Movement of larvae through tissues is certainly made easier by the wood-rotting fungi. The most probable situation is one in which both the fungus and the woody tissues which it permeates each contribute particular nutrients to the larval diet (Buchner, 1965).

Continuity of the symbiotic association into the adult stage of the wood wasp is ensured by means of a complicated process (Parkin, 1941, 1942). Young Siricid larvae are fungus-free but in full-grown female larvae a pair of mycetangia are present in the hypopleural region at the end of the first abdominal segment, hidden in the deep intersegmental fold (Figure 41a, b). Each mycetangium is made up of a series of deep pits in the thickened cuticle, the walls of which bear spines that retain, within each pit, a mass of intertwined hyphae or oidia lying in an oily fluid (Figure 42a, b). The hypopleural cells beneath the mycetangium are enlarged to form a columnar, glandular epithelium. In *Tremex columba* two pairs of mycetangia are present in late-stage larvae (Stillwell, 1965). The first pair lies between the metathorax and the first abdominal segment, the second between the first and second abdominal segments. Shortly before the final larval ecdysis, the epithelial cells secrete copiously, and these secretions harden to form plates containing fungal cells. The plates are shed at the moult and subsequently, when the adult female bites its way from the pupal chamber, fragments of plate material are drawn into the ovipositor pouches by continuous retractive movements of the sting components. The oidia are then released from the plate matrix and begin to grow within the pouches (Francke-Grosmann, 1967).

Symbiotic wood wasp fungi can be easily isolated from mycetangia, egg surfaces, or larval tunnels, and will grow well in axenic culture but identification of them presents some difficulties. Most have been found to be Basidiomycotina and have been variously placed in *Amylostereum, Stereum* or *Daedalea*. There is also an unconfirmed report that the pyrenomycetous species *Daldinia concentrica* may be symbiotic (Cartwright, 1938; Parkin, 1942; Stillwell, 1964; Talbot, 1964; Vaartaja

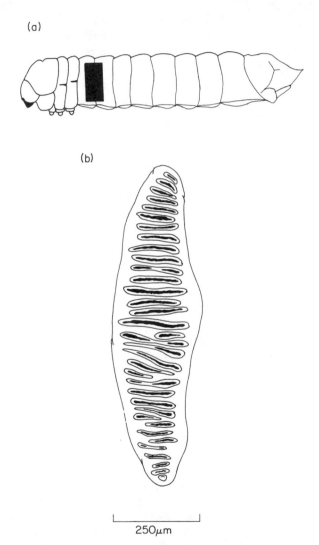

(a)

(b)

250μm

Figure 41   Wood wasp fungi: (a) diagram of lateral view of a wood wasp larva showing position of mycetangium; (b) surface view of mycetangium, fungus-containing pits showing through the integument. (b) Redrawn from Francke-Grosmann, 1967; copyright Academic Press Inc.

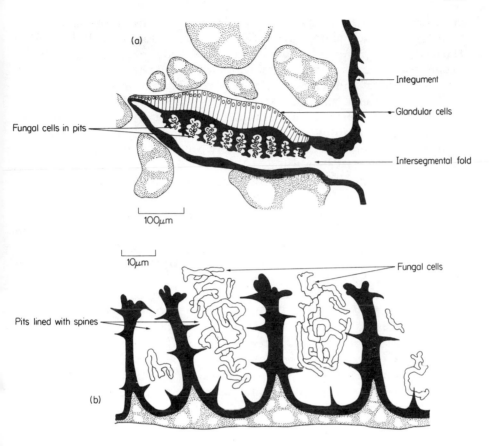

Figure 42   Wood wasp fungi: (a) longitudinal section through the mycetangium of female larva of *Sirex cyaneus*; (b) details of female larval mycetangium of *S. gigas*. Redrawn from Buchner, 1965; by permission of John Wiley & Sons

and King, 1964). While the advantage which the wood wasp larva gains from the symbiosis, if not absolutely clear, may be guessed at with some degree of confidence, that accruing to the fungus is rather more doubtful. The only apparent benefit is that it is deeply inoculated into wood, so that any possible competition from other wood-attacking fungi is delayed until it is well established in the tissues.

2. Food Fungi of Ants

Some termites maintain fungi within their nests but it seems that these have little food value, particularly for large colonies of the insects. The fungi are certainly eaten, although this occurs relatively infrequently, and they perhaps supply vitamins or other growth factors rather than major energy-rich compounds (Grassé, 1945; Lüscher, 1949, 1951; Hartzell, 1967). An additional function of these fungi might be to produce and maintain a favourable microclimate within the nest, where a high humidity and temperature are necessary. In contrast, species of a tribe of myrmicine ants, the Attini, culture saprotrophic fungi which in nature provide their sole and complete diet. The symbiosis is unique in that the ants employ a complicated procedure of husbandry to maintain and feed their fungi in specially prepared subterranean gardens within their nests (Weber, 1966, 1972a, b). The queens and their broods live in these gardens where the fungi are grown on either leaf fragments, which are cut and transported to the nests, or on particles of organic matter mixed with insect excrement. Some of these ants are economically important because of their destruction of crop plants in the search for leaf tissue, while in tropical rain forests they may be of great ecological value through their positive effect on organic matter enrichment of the soil.

For all attine species the development and care of the gardens follows a broadly similar pattern, but the best developed expression of the symbiosis is found in the genus *Atta*. Nests of *Atta* species are founded by fertile females after a nuptual flight, and each carries a small pellet of the symbiotic fungus in an infrabuccal pocket at the rear of its mouth (Figure 43). After excavating a small chamber the queen lays eggs which are of two kinds, alimentary eggs that she herself

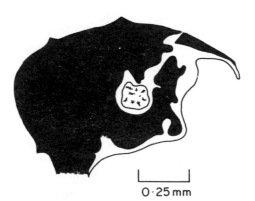

0·25 mm

Figure 43 Longitudinal   section through head of fertile *Atta* female, infrabuccal pocket containing a pellet of viable fungal hyphae. Redrawn from Weber, 1966 after Huber, 1905; copyright 1966 by the American Association for the Advancement of Science

Figure 44   Fungus-growing ants. Queen of *Atta sexdens* with her brood and attendant workers on a young fungus garden. From Weber, 1972; by permission of the American Philosophical Society

consumes and feeds to the first larvae, and reproductive eggs which produce larvae. Several months elapse between the laying of the first eggs and the emergence of the first workers. The queen, meanwhile, has planted the fungus and continually anoints it with anal droplets, feeling its mycelium almost continually with her antennae. She cleans the latter by passing them through a strigil or comb borne on each of her front legs. After emergence the workers then begin to tend and maintain the fungus (Figure 44).

Species of *Atta* and *Acromyrmex*, but not other species, are strongly polymorphic, the size range of workers being continuous from the smallest, or minima caste, through the media castes to the largest, or maxima (Figure 45a–c). The maxima cut leaves and protect the colony, while the media cut leaves but also tend the gardens and feed the brood. The minima are restricted to the gardens where their small size allows them to tend the fungus and to care for the eggs and smaller larvae. The mouthparts of the minima and media ants are modified for their various tasks, particularly for grooming and feeding the fungus, and all have an infrabuccal pocket to receive debris from these activities. During the first few months of the life of a colony only minima or media workers are produced. Leaf fragments are brought to the garden and are cut into pieces 1–2 mm across. The ants lick the fragments and press the margins of them with their mandibles so that the tissues are pulped. The fragments are then treated with anal

112

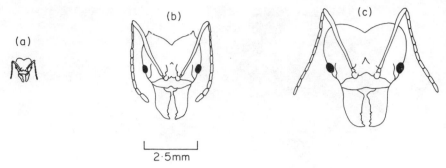

Figure 45   *Atta cephalotes*, heads of worker castes: (a) minima; (b) media; (c) maxima. Redrawn from Weber, 1966; copyright 1966 by the American Association for the Advancement of Science

droplets, are inserted into the garden and are inoculated with mycelial tufts picked up by the ants. The condition of the fungus is continually tested by the ants, who feel it with their antennae, and it is frequently anointed with anal droplets. Hyphae commonly terminate in swollen, cytoplasm-rich organs, called variously gonglydia, bromatia or staphyla, that are cropped and eaten by adult ants and are also fed to the larvae (Figure 46a, b). It used to be thought that these structures were formed in response to the activities of the ants but this is not so and they will form in their absence (Weber, 1966).

Those ant fungi that have so far been identified are Basidiomycotina, for example species of *Auricularia*, *Agaricus* and *Lepiota*, together with,

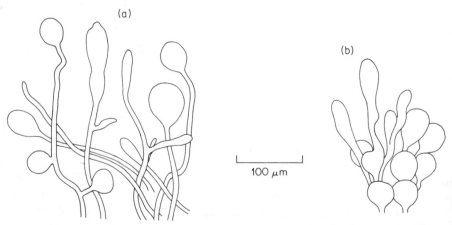

Figure 46   Fungi from ant gardens: (a) staphyla from garden of *Trachymyrmex jamaicensis*; (b) staphyla from garden of *T. septentrionalis*. Redrawn from Weber, 1966; copyright 1966 by the American Association for the Advancement of Science

possibly, the pyrenomycetous genera *Daldinia* and *Xylaria*. No ant fungi can apparently exist in a free-living state outside the gardens so that their association with the ants is probably obligate. They can, however, be grown in axenic culture in which they occasionally produce their sporophores (Weber, 1956a, b, 1957). Within the gardens of any ant species, growth of only one species of fungus takes place, even though conditions in the gardens are suitable for the development of other fungi, and alien spores must be brought in continually on tissue fragments or the bodies of workers. The maintenance by tne ants of a luxuriant fungal monoculture in non-sterile conditions is remarkable, but how they achieve this is not clear. Foreign moulds may be physically weeded out to some extent but their eradication by this method is obviously impossible. Fungistatic and bacteriostatic factors may be present in anal liquid or saliva which prevent growth of contaminants but the evidence for these is conflicting. *Trachymyrmex septentrionalis* will maintain a fungus garden on agar. When contaminating fungi and bacteria appear on the agar, pieces of the contaminated substrate are cut out and piled some distance from the symbiotic fungus. Growth of fungi and bacteria on these pieces, which have been coated with saliva during their removal, appears to be retarded (Weber, 1956a, 1966). Another attine, *Atta columbica tonsipes*, has been examined for antibiotic secretion with negative results (Martin, MacConnell and Gale, 1969). Some species do, however, secrete phenyl acetic acid from their metapleural glands and this compound has antibiotic effects on at least some bacteria and fungi (Maschwitz, Koob and Schildkneckt, 1970). The degree of dependence of an ant species on a particular species of fungus has not been determined. If gardens are exchanged between ant species then the ants reject them, but this could be due to the presence of pheromones from the original ant partner. If axenic cultures of the fungi are exchanged, the ants do not reject these and each species will feed, albeit to a limited extent, on the food fungus of the other (Weber, 1957).

The symbiotic fungi utilize the cellulose in the plant tissues supplied to them by the ants and provide the insects with a complete carbohydrate- and nitrogen-rich diet in the form of fungal cytoplasm and vacuolar sap. The fungi in the form of amino acids in faecal material and MacConnell, 1969; Martin and Weber, 1969). The ants supply nitrogen to the fungi in the form of amino acids in faecal material (Martin and Martin, 1970a). Rectal fluid, in addition to nitrogen compounds, contains proteases, and these release additional nitrogen from the substrate on which the fungi are growing which the latter can then utilize. Secretion of proteases in rectal fluid seems to be restricted to attine ants and may be an important feature of the symbiosis since the fungi do not appear to have a well-developed proteolytic ability. (Martin, 1970; Martin and Martin, 1970b, 1971).

3. Septobasidium and Scale Insects

Species of the Basidiomycotina genus *Septobasidium* are epiphytic, usually on trees, and are invariably associated with scale insects. Growth of the fungus on the plant is entirely superficial, but some at least of its nutrients are obtained from the plant indirectly through exploitation of the insects with which the fungus is symbiotic. *Septobasidium* species are sometimes considered to be endosymbionts but this is not strictly accurate. The bulk of their somatic tissue is outside the insect, the latter being invaded by only a few hyphae that enter its body but do not penetrate its cells.

Aspects of this complex, fungus—insect—tree relationship are well known but few experimental studies of it have been made (Couch, 1938; Watson, Underwood and Reid, 1960). Consequently, many details of the physiology of the relationship can only be guessed at. *S. burtii*, which is epiphytic on oaks and fruit trees, is probably the most studied species (Couch, 1938). Its strongly-ridged, resupinate, lichen-like thalli are perennial and may be up to several centimetres in diameter. Within them is a labyrinth of chambers and tunnels containing individuals of the scale insect *Aspidiotus osborni*. Tunnels open to the surface by means of slit-shaped entrances, while chambers usually contain a single insect which lies in direct contact with the bark of the tree.

In the Spring, basidiospores, which are capable of budding, are formed on the thallus surface, and sporulation coincides with the

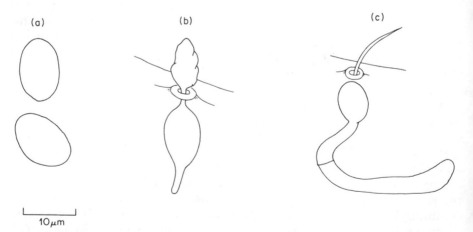

(a) (b) (c)

10μm

Figure 47 *Septobasidium burtii*, stages in infection of the scale insect *Aspidiotus osborni*: (a) ungerminated bud cells; (b) germ tube passing through setal pore and giving rise to infection cell; (c) infection cell beneath setal pore producing a septate hypha. Redrawn from Couch, 1938; by permission of the University of North Carolina Press

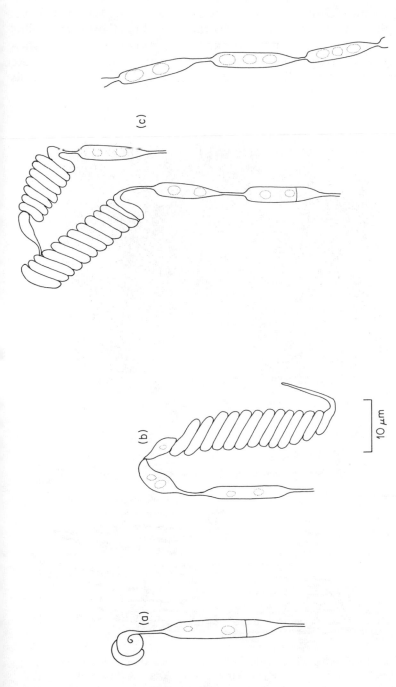

Figure 48 S. burtii, development of mycelium within A. osborni: (a) and (b) spindle-shaped cells giving rise to hyphals coils; (c) spindle-shaped cells linked by fine hyphae. Redrawn from Couch, 1938; by permission of the University of North Carolina Press

viviparous production of young by fertile females within the thalli. Young insects may either stay where they are born, within the fungus, or crawl to the outside. Infected insects then crawl back into the thallus or settle down on uncolonized bark. Those re-entering the thallus become responsible for its continued survival and growth, while those settling upon clean bark distribute the fungus to fresh sites. Bud-cells

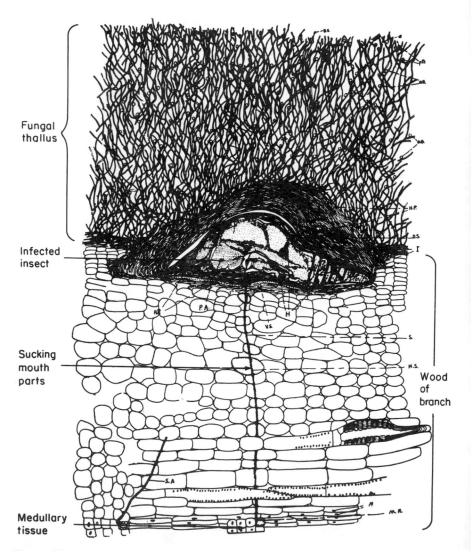

Figure 49   *S. burtii*, infected individual of *A. osborni* within its fungal chamber on the bark of a branch. Long, sucking mouthparts are inserted into the medullary rays of the tree. From Couch, 1938; by permission of the University of North Carolina Press

adhere to the legs, antennae, or body surface and short hyphae arise from them which pass through the integument either directly or by means of the setal pores (Figure 47). After entry each hypha immediately swells to produce a large infection cell within the insect's body. Growth of the fungus then proceeds by alternate formation of short, fine hyphae and fatter, spindle-shaped cells, so that a mycelium comprising chains of cells is formed within the haemocoel (Figure 48a, b). Several weeks after initial infection, symmetrically-coiled assimilative hyphae are produced on the internal mycelium and, simultaneous with the appearance of these, fine hyphae pass out of the setal pores to the outside of the insect (Figure 48c). The end of each fine, external hypha enlarges into a spherical cell which gives rise to a further, coiled hypha.

If the infected insect has returned to the thallus it becomes surrounded, but not penetrated, by hyphae which grow from the tunnel or chamber floor. These anastomose with the coiled hyphae emerging from the insect's body. If the infected insect has moved to fresh bark the hyphae emerging from it grow and branch to form a small thallus surrounding the body. Other infected insects may settle down in contact with or beneath this. The mouth and spiracles of infected insects do not usually become covered by hyphae and the insect is free to move in a limited way within its chamber. Its sucking mouthparts pierce the bark and it feeds, apparently quite normally, on the contents of medullary ray cells (Figure 49). It has been suggested that only females develop to maturity within thalli, that only females become infected, and that males mature free-living on the bark (Couch, 1938).

Effects of S. burtii on the Insects

The fungus has two components, one outside and one inside the insect, and these relate to the insect in different ways. Scale insects can live unassociated with the fungus but they probably do so rarely and the asymbiotic life might be a difficult one. The thallus protects both infected and uninfected insects from unfavourable environmental extremes and guards them against hymenopterous wasps. This protection is afforded at the expense of the infected members of the insect population, the latter losing their mobility, remaining dwarfed in size, and being rendered infertile. Their longevity may, however, be increased. There is therefore a mixture of antagonistic and mutualistic effects with the balance on the side of benefit to the insects.

Although there is some evidence of host defence reactions to invasion through phagocytosis of hyphal coils, clearly a high degree of tolerance to penetration by the fungus has been developed by the insect. Even so, hyphae and assimilative coils are not intracellular and are restricted to the haemocoel.

Effects of the Insects on S. burtii

As well as acting as the sole agents of dispersal of the fungus, infected insects feed continuously on medullarly ray cells. Nutrients are then assimilated from their blood by the hyphal coils and these nutrients are translocated within the thallus to its growing margin. Thallus diameter and rate of growth seem to be related to the number of infected insects living within it. During the summer the ratio of infected to uninfected individuals within the thallus increases, while at the approach of winter the ratio gradually decreases. How this regulation is achieved is not known.

Also unexplored are those physiological problems which must be an inevitable consequence of a fungus using an insect as a nutrient bridge between itself and a higher plant. There must be a number of important changes in the metabolism of both insect and plant, and the development of mechanisms to maintain one-way movement of nutrients to the fungus via the insect.

The fungus—insect symbiosis allows each organism to occupy a habitat that neither could colonize independently, but the degree of dependence of the fungus upon the insect is perhaps greater than that of the insect upon the fungus, since if insects are removed from thalli the latter stop growing and die. Although S. burtii can be grown in axenic culture, where it does not sporulate, in nature it is obligately associated with scale insects.

Effects of the Fungus—Insect Symbiosis on the Plant

Depending upon the species, Septobasidium may or may not damage the plant on which it grows. Where serious damage does not occur there may, nevertheless, be localized killing of the bark or staining and erosion of the underlying wood. Since there is no evidence to suggest deep fungal penetration, it is probable that these effects are due to the activity of the feeding insects (Watson, Underwood and Reid, 1960). There is, though, a possibility that extracellular fungal enzymes or metabolities might be involved. Severe wood injury leading to death of part or the whole of the tree does sometimes occur but it is not always clear whether this is attributable to the activities of the fungus alone, the insect alone, or both acting together. Slight hyphal penetration of bark has been noted, as have abnormal growth responses of branches bearing heavy thallus growth. In some Septobasidium associations the fungus—insect relationship may change suddenly. There can be unaccountable increases in nutrient demand by the fungus leading to death of the insects. The fungus may then become more active in attacking the tree.

Septobasidium species are unique in that, while they are biotrophic in the sense that they require association with living host tissues, they never occupy living host cells but develop their absorptive structures within the host's haemocoel. There is also the possibility that, simultaneous with this biotrophy, parts of the thallus may be saprotrophic, utilizing dissolved nutrients on the surface of the tree bark.

4. Mycetome Fungi

Many insect species possess endosymbiotic, biotrophic yeasts or yeast-like fungi which, although some can be grown in axenic culture, appear to be ecologically obligately symbiotic. The fungi may grow in the lymph between fat body cells or, more commonly, occupy living host cells that are specially adapted to receive them. These cells, mycetocytes, are usually aggregated to form a distict organ, the mycetome. Some insects house only fungi, others contain exclusively bacteria or other unidentified symbiotic micro-organisms, while in some instances plurisymbioses occur where bacteria, fungi, and other organisms share mycetocytes within the mycetome.

Mycetome fungi occur mainly in the Homoptera and Coleoptera although other insect orders, so far not explored in detail, may prove to contain them (Koch, 1960, 1967; Buchner, 1965). Within an insect order, the possession of biotrophic endosymbionts by particular genera is, in general, related to feeding habit. For example, eaters of dead organic matter, or insects that suck plant sap have a low nitrogen, and perhaps also a vitamin-deficient, diet. Many endosymbionts have been shown to make good these deficiencies for the host, allowing the utilization of foods and the occupation of habitats that might be otherwise nutritionally inadequate (Richards and Brooks, 1958; Brooks, 1963).

It is difficult to visualize all the benefits that might accrue to an endosymbiotic fungus from its association with an insect. The only immediately apparent one is that host cells may provide a congenial environment for its development if the fungus has poor saprotrophic ability, or cannot withstand other kinds of competition from other micro-organisms. The various transmission mechanisms that have been developed by insects for transferring fungal cells from one host to another allow continuity of fungal growth in successive host generations without the interpolation of a free-living saprotrophic or dormant phase. The tolerance of host cytoplasm to occupation by fungal cells, and the probable resistance of the latter to the effects of host enzymes, emphasizes the highly-developed association between host and symbiont. It also implies that in evolutionary terms these associations are very ancient.

Location and Transmission

Structural and behavioural aspects of the symbioses have great diversity and this, together with a frequent lack of experimental studies, particularly physiological investigations, makes a general account difficult. The most comprehensive studies have been made on species within two beetle families, the Anobiidae and the Cerambycidae.

In all symbiotic Anobiids the mycetomes are located at the point where fore- and midgut meet. Here large, subdivided caeca are formed from evaginations of the midgut, and within them the endosymbiont is contained (Figure 50a, b). The epithelium of these intestinal myce-

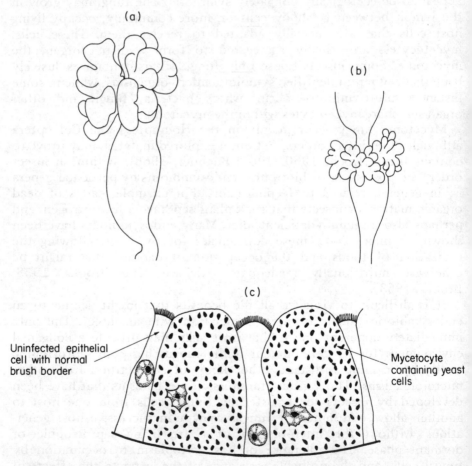

(a)

(b)

(c)

Uninfected epithelial cell with normal brush border

Mycetocyte containing yeast cells

Figure 50 *Sitodrepa panicea*: (a) and (b) mycetomes of larva and adult respectively formed by evagination of the midgut; (c) details of mycetome epithelium. (a) and (b) Redrawn from Buchner, 1965 after Koch; by permission of A. Koch and John Wiley & Sons. (c) Redrawn after Breitsprecher, 1928

tomes is composed mainly of mycetocytes which are enlarged, lack a brush border, are packed with budding fungal cells and have stellate nuclei (Figure 50c). Mycetocytes lack the ability to divide and may be polyploid. Interspersed between the mycetocytes are normal epithelial cells with a typical brush border. Some of the mycetocytes always contain a number of dead fungal cells and this fungal material is usually continually expelled into the mycetome lumen. During metamorphosis the midgut epithelium breaks down and fragments of epithelium, fungal cells, and entire mycetocytes are liberated into the lumen of the intestine. The shrivelled mycetomes become engulfed by the developing imaginal intestine and viable fungal cells then infect the new intestinal cells and multiply within them.

Continuity of the symbiosis between adult and larva is ensured by means of additional fungus-bearing organs. These are intersegmental tubules that are continuous with the ovipositor sheath (Figure 51).

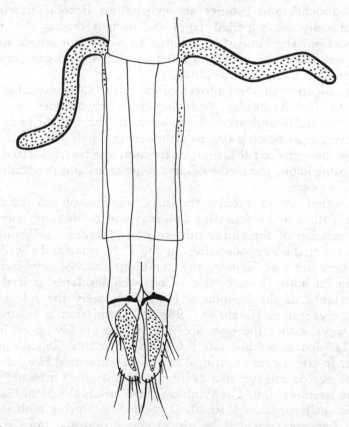

Figure 51   *Sitodrepa panicea*, egg-smearing organ with fungus-filled intersegmental tubules and vaginal pockets. Redrawn after Breitsprecher, 1928

Glandular cells lining them produce a secretion within which fungal cells are suspended and grow, so that when an egg passes through the ovipositor it draws fungal cells down with it. There are also two fungus-filled pouches at the end of the ovipositor under the vagina. As the egg passes from the vagina, additional fungal cells from these pouches are smeared upon it. Larvae become infected at hatching when they eat their contaminated egg cases, and yeast cells pass undigested to the intestinal epithelium which they then invade. The mycetomes are already partly formed and are ready to receive the symbiont.

Yeast cells ingested by the newly-hatched larvae lodge in the epithelium of the mycetome primordia and begin to bud. In some insect species cytoplasmic processes push out from the mycetocytes and envelop yeast cells. Mycetocytes lacking a brush border are pre-formed within mycetangial primordia in the absence of the endosymbiont within the gut lumen. In the adult, the intersegmental tubules act as reservoirs to replenish the vaginal pouches. In the pupa, both pouches and tubules are empty but become inoculated when fungal cells are expelled from the midgut during the imago's first defaecation. The hindgut opens into the ovipositor sheath so that fungal cells can reach the secretion-filled transmission organs and they then grow within them saprotrophically.

The location of the endosymbiont in the Cerambycidae is similar to that in the Anobiidae, but the larval mycetomes are lost during metamorphosis and are not replaced. In the female imago there are, however, egg-smearing organs although the male is symbiont-free. This change in symbiont location, or its complete loss, is related to a change in feeding habit, the larvae being xylophagous and the adults feeding on nectar or sap.

In other insect groups the mycetome is by no means always a modification of the intestine and may not be directly connected to it. Transmission of the endosymbiont must therefore be brought about in ways other than egg-smearing. In the Fulgoridae and Cicadinae, whose members are sap feeders, the symbiont may occupy individual cells within fat body tissue or live within well-developed paired or unpaired mycetomes in the abdomen. Fungi are rarely the sole occupants of these mycetomes (Buchner, 1965). Transmission is ensured in one of two ways, either the eggs are infected before laying or, in viviparous insects such as aphids, late infection of well-developed embryos takes place. In either case the fungal cells are transported from the mycetome to the egg or embryo tissues by the migration of infected host cells or in the haemolymph. The symbiont then enters the germ tissues through special cells produced for their reception. During embryogenesis the fungi become relacated in mycetome primordia and, in the case of plurisymbioses, different symbionts may become relocated in different parts of the mycetome. Both symbiont movement and multiplication are under very fine control.

Nature of the Symbiosis

It has not always proved possible to grow axenic cultures of mycetome fungi, particularly those involved in plurisymbioses, and even where cultures have been obtained difficulties have been encountered in positive identification of the fungi. Endosymbionts of the Anobiidae have been placed in *Torulopsis, Taphrina* and *Symbiotaphrina*, while those from the Cerambycidae seem to be anascosporogenous yeasts resembling *Candida* and produce pseudomycelia in culture. Details of the physiological consequences of infection by symbionts have been obtained almost exclusively from studies on *Sitodrepa panicea*, the drug store beetle (Foekler, 1961; Jurzitza, 1962; Kühlwein and Jurzitza, 1960, 1961; Jurzitza, 1964). It is, therefore, not possible to make wide generalizations with any degree of certaintly. Host—symbiont physiology has been investigated using three main methods; by studying the consequences of symbiont loss by the insect, by examining the nutritional characteristics of the symbionts in axenic culture, and by attempting artificial infection of asymbiotic hosts.

With *Sitodrepa panicea*, egg surface sterilization followed by aseptic rearing techniques have been used to produce asymbiotic larvae, and it has been demonstrated that symbiont loss has a number of far-reaching effects. There is inevitably retardation of growth, an increase in death rate, and difficulty during moulting, with the adult stage rarely being reached. It is almost certain that normal development of symbiotic larvae is not due to the direct influence of the fungi on endocrine systems, but that these systems are affected indirectly through the supplementation of marginal or submarginal diets by the endosymbionts (Richards and Brooks, 1958). The deleterious consequences of asymbiosis can to various degrees be alleviated by the provision of B-group vitamins and sterols in the diet (Blewett and Fraenkel, 1944). The fungi can synthesize at least some B-group vitamins, but there is no direct evidence that they produce sterols. Not all endosymbionts are, however, autotrophic for vitamins and some themselves have partial requirements for biotin or thiamine. In addition to providing vitamins, and possibly sterols, the endosymbionts may also play an important role in their host's nitrogen metabolism. For instance, symbiotic larvae of *S. panicea* require only arginine, leucine and threonine in their diet for normal development. Asymbiotic larvae require all essential amino acids *plus* glycine and tryptophan (Pant, Gupta and Nayar, 1960). Studies on axenic cultures of Anobiid fungi show that some can utilize uric acid and urea as nitrogen sources, and it has been suggested that they can utilize their host's waste products for amino acid and protein synthesis (Jurzitza, 1959). Further involvement in nitrogen metabolism could occur if the endosymbionts were capable of fixing free nitrogen. Many bacterial symbionts in mycetomes presumably do this but there is good reason to doubt that any fungi have such an ability (Peklo and Satava, 1950; Metcalfe and Chayen, 1954; Millbank, 1969).

One unique effect of fungal infection has been recorded in the Coccid species *Stictococcus diversisetae* (Buchner, 1954, 1955). Fungal cells carried with host cells invade the ovary but infect only those öocytes beside which they come to lie. This chance infection produces two kinds of egg, infected and uninfected. Both develop partheno-gentically but infected eggs give rise to females while uninfected eggs produce males. Production of males involves loss of a chromosome and it appears that the presence of the endosymbiont prevents this loss.

To what degree mycetome fungi are host-specific is not entirely clear. A number of attempts have been made to artificially infect one host species with a fungus from a different species, or to infect hosts with non-symbiotic yeasts (Fraenkel, 1952; Foekler, 1961). Symbiont exchange has been achieved between two Anobiids, *Sitodrepa panicea* and *Lasioderma serricorne*, and their mycetomes seem to remain quite normal. Although the *Lasioderma* symbiont can fulfil the function of the *Sitodrepa* symbiont the reverse is not the case, the symbiotic larvae still requiring thiamine in their diet. This indicates that this particular endosymbiont may neither fully satisfy its natural host's thiamine requirement nor that of its adopted host.

Most attempts at inoculation of insects with non-symbiotic yeasts have been unsuccessful although the yeast cells are not digested within the gut and remain viable. The single exception is the fodder yeast *Torulopsis utilis* which will infect *S. panicea*. The epithelial cells of the mycetome become infected but still retain their normal brush border. In addition, all cells of the midgut epithelium, including those that are not normally colonized by symbiotic yeasts, are invaded. During metamorphosis *T. utilis* is eliminated from the gut, and the adult gut and transmission organs subsequently remain fungus-free. If a mixture of *T. utilis* and the *Sitodrepa* symbiont are fed to sterile larvae, the symbiont is preferentially taken up by the epithelial cells, and only the mycetome epithelium is then invaded. The presence of the symbiont within the mycetomes must change the ability of the remainder of the midgut epithelium to take up the foreign yeast.

Bibliography

Baker, J. M., 1963. In P. S. Nutman and B. Mosse (eds.), *Symbiotic Associations*, p. 232–265. 13th *Symp. Soc. gen. Microbiol.*, Cambridge University Press.
Baker, J. M. and Kreger-van Rij, N. J. W., 1964. *Antonie van Leeuwenhoek*, 30: 433–441.

Baker, J. M. and Norris, D. M., 1968. *J. Invert. Path.*, 11: 246—250.
Batra, L. R., 1963. *Am. J. Bot.*, 50: 481—487.
Batra, L. R., 1967. *Mycologia*, 59: 976—1017.
Batra, L. R. and Francke-Grosmann, H., 1961. *Am. J. Bot.*, 42: 453—456.
Batra, L. R. and Francke-Grosmann, H., 1964. *Mycologia*, 56: 632—636.
Blewett, M. and Fraenkel, G., 1944. *Proc. R. Soc. B.*, 132: 212—221.
Breitsprecher, E., 1928. *Z. Morphol. Ökol. Tiere*, 11.
Brooks, M. A., 1963. In P. S. Nutman and B. Mosse (eds.), *Symbiotic Associations*, p. 200—231. 13th *Symp. Soc. gen. Microbiol.*, Cambridge University Press.
Buchanan, W. D., 1941. *J. econ. Ent.*, 34: 367.
Buchner, P., 1954. *Z. Morph. Ökol. Tiere*, 43: 262—312.
Buchner, P., 1955. *Z. Morph. Ökol. Tiere*, 43: 397—424.
Buchner, P., 1965. *Endosymbiosis of Animals with Plant Micro-organisms*, John Wiley, New York and London.
Cartwright, K. S. G., 1938. *Ann. appl. Biol.*, 25: 430—432.
Chu, H. M., Norris, D. M. and Kok, L. T., 1970. *J. Insect Physiol.*, 16: 1379—1387.
Couch, J. N., 1938. *The Genus Septobasidium*, University of North Carolina Press.
Foekler, F., 1961. *Z. Morph. Ökol. Tiere*, 49: 78—146.
Fraenkel, G., 1952. *Tijdschr. Ent.*, 95: 183—196.
Francke-Grosmann, H., 1967. In S. M. Henry (ed.), *Symbiosis*, Vol. 2, p. 141—205, Academic Press, New York.
Graham, K., 1967. *A. Rev. Ent.*, 12: 105—126.
Grassé, P. P., 1945. *Annls. Sci. nat. (Zool.)*, 7: 115—146.
Happ, G. M., Happ, C. M. and Barras, S. J., 1971. *Tissue Cell*, 3: 295—308.
Hartzell, A., 1967. In S. M. Henry (ed.), *Symbiosis*, Vol. 2, p. 107—140, Academic Press, New York.
Huber, I., 1905. *Biol. Centralbl.*, 25: 606.
Jurzitza, G., 1959. *Arch. Mikrobiol.*, 33: 305—332.
Jurzitza, G., 1962. *Arch. Mikrobiol.*, 43: 412—424.
Jurzitza, G., 1964. *Arch. Mikrobiol.*, 49: 1—22.
Koch, A., 1960. *A. Rev. Microbiol.*, 14: 121—140.
Koch, A., 1967. In S. M. Henry (ed.), *Symbiosis*, Vol. 2, p. 1—106, Academic Press, New York.
Kok, L. T., Norris, D. M. and Chu, H. M., 1970. *Nature, Lond.*, 225: 661—662.
Kühlwein, H. and Jurzitza, G., 1960. *Naturwissenschaften*, 47: 547.
Kühlwein, H. and Jurzitza, G., 1961. *Arch. Mikrobiol.*, 40: 247—260.
Leach, J. G., Hodson, A. C., Chilton, S. J. P. and Christensen, C. M., 1940. *Phytopathology*, 30: 227—236.
Lowe, R. E., Giese, R. L. and McManus M. L., 1967. J. Invert. Path., 9: 451—458.
Lüscher, M., 1949. *Acta Trop.*, 6: 161—165.
Lüscher, M., 1951. *Nature, Lond.*, 167: 34—35.
Martin, M. M., 1970. *Science N.Y.*, 169: 16—20.
Martin, M. M. and Martin, J. S., 1970a. *J. Insect Physiol.*, 16: 109—119.
Martin, J. S. and Martin, M. M., 1970b. *J. Insect Physiol.*, 16: 227—232.
Martin, M. M. and Martin, J. S., 1971. *J. Insect Physiol.*, 17: 1897—1906.
Martin, M. M., Carman, R. M. and MacConnell, J. G., 1969. *Ann. ent. Soc. Am.*, 62: 11—13.
Martin, M. M., MacConnell, J. G. and Gale, G. R., 1969. *Ann. ent. Soc. Am.*, 62: 386—388.
Martin, M. M. and Weber, N. A., 1969. *Ann. ent. Soc. Am.*, 62: 1386—1387.
Maschwitz, U., Koob, K. and Schildkneckt, H., 1970. *J. Insect Physiol.*, 16: 387—404.
Mathieson-Käärik, A., 1960. *Oikos*, 11: 1—25.
Metcalfe, G. and Chayen, S., 1954. *Nature, Lond.*, 174: 841—842.

Millbank, J. W., 1969. *Arch. Mikrobiol.*, 68: 32—39.

Norris, D. M. and Baker, J. M., 1967. *Science N.Y.*, 156: 1120—1122.

Pant, N. C., Gupta, P. and Nayar, J. K., 1960. *Experientia*, 16: 311—312.

Parkin, E. A., 1941. *Nature, Lond.*, 147: 329.

Parkin, E. A., 1942. *Ann. appl. Biol.*, 27: 268—274.

Peklo, J. and Satava, J., 1950. *Experientia*, 6: 190—192.

Richards, A. G. and Brooks, M. A., 1958. *A. Rev. Ent.*, 3: 37—56.

Saunders, J. L. and Knoke, J. K., 1967. *Science N.Y.*, 157: 460—463.

Schedl, K. E., 1958. *Proc. 10th Int. Congr. Ent. Montreal. (1956)*, 1: 185—197.

Shifrine, M. and Phaff, H. J., 1956. *Mycologia*, 48: 41—55.

Stillwell, M. A., 1964. *Can. J. Bot.*, 42: 495—497.

Stillwell, M. A., 1965. *Can. Ent.*, 97: 783—784.

Talbot, P. H. B., 1964. *Aust. J. Bot.*, 12: 46—52.

Vaartaja, O. and King, F., 1964. *Phytopathology*, 54: 1031—1032.

Watson, W. Y., Underwood, G. R. and Reid, J., 1960. *Can. Ent.*, 92: 662—667.

Weber, N. A., 1956a. *Ecology*, 37: 150—161.

Weber, N. A., 1956b. *Ecology*, 37: 197—199.

Weber, N. A., 1957. *Ecology*, 38: 480—494.

Weber, N. A., 1966. *Science N.Y.*, 153: 587—604.

Weber, N. A., 1972a. *Gardening Ants: the Attini*, American Philosophical Society, Philadelphia.

Weber, N. A., 1972b. *Am. Zool.*, 12: 577—587.

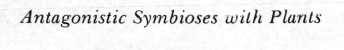

Antagonistic Symbioses with Plants

Chapter 8

Facultative Necrotrophs

A great number of diseases of higher plants are caused by facultatively necrotrophic fungi. These species have a well-developed ability to actively colonize dead organic substrates in the face of competition from other micro-organisms, and a capacity to maintain themselves indefinitely in a saprotrophic condition within such substrates. If a suitable living host is contacted, then the fungi become necrotrophic within it, usually causing rapid destruction of the invaded tissues. Host cells are first killed through the activities of fungal hyphae within the tissues and are then exploited saprotrophically in a manner identical to that in which free-living saprotrophic fungi exploit their substrates. The antagonistic effects of facultative necrotrophs are the result of the continual creation of dead host material within which they can successfully develop as saprotrophs. They may, therefore, be looked upon in some ways as being saprotrophs growing out of place but, unlike free-living obligately saprotrophic fungi, they possess mechanisms by which they can enter living tissues, survive in contact with them, and eventually bring about their death. In some respects they resemble fungi that cause severe and deep-seated diseases in vertebrate animals and, like these, they are easily brought into axenic culture where they have simple nutrient requirements. Although the existence and ubiquity of facultative necrotrophs is indisputable it is not always possible, with any degree of confidence, to place species which apparently have most of the attributes of facultative necrotrophy in this group of fungi. The capacity for indefinite saprotrophy depends on an ability to move from dead colonized organic substrates to other, virgin, dead substrates. Yet for many necrotrophs detailed information as to the extent, efficiency and persistence of saprotrophy is lacking. Many necrotrophic fungi may be isolated with great frequency from soil or decaying organic matter, and the spores of many exist as a normal component of the air spora, but this need not necessarily reflect their success as free-living saprotrophs. Their growth, sporulation and survival may be strictly limited to dead host tissues which they themselves have killed, and they may entirely lack the ability to colonize other dead substrates in the face of competition from other micro-organisms, that is they are ecologically obligately necrotrophic.

These problems are discussed in more detail in the consideration of obligate necrotrophs in Chapter 10.

Typical examples of facultative necrotrophs are species of *Aspergillus, Penicillium* and *Rhizopus*, which have a wide host range and cause extensive breakdown of tuber and fruit tissues, and some fungi that cause damping-off and other seedling diseases. Some species of *Botrytis*, particularly *B. cinerea*, and some strains of *Rhizoctonia solani* probably also belong here, although their status as free-living saprotrophs is not always clear. At least some species or races of *Fusarium* and *Verticillium* which cause rots or vascular-wilt diseases are facultative necrotrophs, while other species of the same genera, though undoubtedly necrotrophs, may be occupiers of dead host tissue rather than active colonizers of dead organic substrates (Garrett, 1970). It might be noted that vascular-wilt fungi go through a distinct, and perhaps unique, pattern of change in their nutritional habit during exploitation of their hosts. They are necrotrophic as they pass through living root tissues into the vascular tract; are then saprotrophic within the vascular elements; are necrotrophic again as they pass out from these as the host becomes moribund; and are finally saprotrophic within dead host tissues. The fungi of orchidaceous mycorrhizas may also occasionally be facultative necrotrophs. These kinds of mycorrhizas have always been looked upon as involving a mutualistic relationship between the host and its mycorrhizal fungus but it is now clear that, rather than mutualism, balanced necrotrophy of the *host* upon the fungus is the basis of this relationship (Lewis, 1973). However, necrotrophy by the fungus within infected host tissues does occasionally occur, which damages the host and may even result in its death (Harley, 1969).

When necrotrophs are active within living host tissues a wide range of effects, from localized necrosis to massive tissue destruction and disintegration result. These effects may extend beyond the physical limit of growth of the necrotrophs and are preceded by, or accompanied by, irreversible changes in the respiratory patterns and permeability of host cells. Changes in metabolism of host cells also occur when necrotrophs are growing within, and are restricted to, functional but non-living tissues, for example when vascular-wilt fungi are developing saprotrophically within the xylem elements.

Details of symbioses involving necrotrophic fungi depend on the particular host—fungus combination being considered, and obviously vary widely from combination to combination, but a number of general features can be distinguished. Some symbioses between necrotrophs and their hosts may be looked upon as having two distinct phases. The first occurs during initial invasion of host tissue and is marked by a transient contact between the hyphae of the fungus and living cells of the host. After death of these cells this phase is succeeded by a second of relatively long duration which is characterized by lack of direct contact between hyphae and living host cells, only killed tissue being occupied. In many necrotrophs the brief initial phase is entirely lacking and host cells are killed under sites of hyphal activity but before hyphae have gained entry to the host. This may occur, for example, during penetration of intact cuticle, wound tissues, or natural discontinuities in the host's surface such as lenticels. Where direct contact with living host cells does occur the latter are rarely, if ever, normal in the sense that their metabolism has probably already been markedly modified by the activity of the necrotroph during the infection process and before actual contact has been made.

Although the necrotroph is living saprotrophically within dead tissues these in turn are contained within living host tissues, the metabolic activities of which range from the apparently normal to the highly abnormal, depending on their distance from the site of fungal activity. The necrotroph is thus surrounded by a series of gradually extending zones of dead, dying and doomed host cells. Within these zones host cells may be variously affected by degradative fungal enzymes, or in some situations by fungal toxins, and by the products of their own abnormal metabolism. The necrotroph in turn may be affected by the products of its own extracellular degradative enzymes, and by a wide range of compounds deposited either within or released from host cells as a result of its necrotrophic and saprotrophic activities. The *milieu* of the fungus is, therefore, physically and biochemically exceedingly complex and its nature is being continually modified as the fungus grows within the host. In some situations such modifications may be inimical to fungal growth, and in these cases activity of the necrotroph may be retarded or severely restricted. Against this background of complexity it is frequently difficult, if not impossible, to distinguish those changes in host tissue that are due to the direct effects of the fungus on living host cells from those that result from the action upon those cells of metabolites released during saprotrophic growth. Moreover, the release from the fungus of metabolites which drastically affect host physiology does not necessarily imply that the resultant effects, while being a characteristic feature of the symbiosis, are an *essential* feature of it, or that they are produced by fungal compounds released or 'designed' for that specific purpose. These effects may merely be an inevitable consequence of a particular kind of saprotroph growing in indirect contact with living host tissues. In view of this, discussion of these fungi is confined to very restricted aspects of their interactions with their hosts.

1. Changes in the Physiology of Invaded Tissues

Entry of a necrotroph into a susceptible host is marked by an irreversible increase in the permeability of host cells in the invaded tissues and a consequent rise in the oxygen consumption of those tissues. Such changes may also occur in host tissues at some distance from the site of fungal activity (Thatcher, 1939; Wheeler and Hanchey, 1968). These effects are brought about in a number of ways. Direct interference by the fungus with the physical integrity of host cells, the action of cellulolytic and pectolytic enzymes and of a wide variety of toxins on host membranes have all, either singly or in combination, been shown to induce electrolyte loss from host tissues (Lai, Weinhold and Hancock, 1968; Calonge, Fielding, Byrde and Akinrefon, 1969; Mount, Bateman and Grant-Basham, 1970; Wood, Ballio and Graniti, 1972; Strobel, 1974). Severe trauma to membrane integrity, uncontrolled respiration and other simultaneous or concatenate events quickly lead to death of host protoplasts. However, in some instances this apparently straightforward sequence of events may mask a more complicated situation. This is particularly true of the nature of the relationship between pectolytic enzyme action and cell death.

Many diseases caused by necrotrophic fungi are characterized by rapid maceration of parenchymatous tissues so that the latter become watery in consistency, due to leakage of vacuolar sap through the dead cytoplasm, and cells separate easily under slight mechanical pressure. Protoplasts are usually killed soon after the tissue begins to lose its cohesion but there is no evidence to suggest that toxins are involved. There is, however, a close relationship between the rate of host cell wall breakdown and rate of cell injury, as indicated by electrolyte loss, and there are strong indications that cell death is caused by the action of pectolytic enzymes bringing about a lethal increase in the permeability of host membranes (Hall and Wood, 1970). Chain-splitting pectolytic enzymes are highly substrate specific, and within the host their substrates are mainly confined to the middle lamella and the pectin-rich matrix of the primary cell wall. The secondary cell wall, which abuts directly on the plasmalemma, contains much lower concentrations of suitable substrates. Since direct action of pectolytic enzymes on the material of the plasmalemma seems highly unlikely, the killing effect of enymes would seem to depend on their altering the physical or physiological relationship between this membrane and the inner layers of the cell wall matrix (Hall and Wood 1973). Little is known concerning plasmalemma-cell wall relationships, but possible ex-planations have been put forward as to how enzyme action might indirectly affect plasmalemma structure and function. Pectolytic activity in the middle lamella and the adjacent cell wall causes the cementing gel to become liquid so that the plasmodesmata and that part of the plasmalemma which covers them become exposed to this solution. If the solution is hypertonic then the exposed regions of the plasmalemma could be ruptured by osmotic forces (Hall and Wood, 1970; 1973). However, in a number of situations, for example during experimental plasmolysis of plant tissues or during the preparation of epidermal strips, plasmodesmata are clearly ruptured but this does not lead to death, the membrane presumably being retracted and reformed within the cell. A curious feature of the effect of pectolytic enzymes on protoplasts is that, at least in some situations, if host cells are plasmolysed before being treated with enzyme preparations then permeability changes and death are delayed, but the cell separation process remains unaffected (Hall and Wood, 1974). This indicates that the plasmalemma is in some way protected from the indirect effects of middle lamella and cell wall degradation. Studies on the effects of an endopectate lyase on potato tuber tissues have shown that, if protoplasts which have been protected in this way are deplasmolysed in the absence of the enzyme, then death of the protoplast is an immediate result (Basham and Bateman, 1975). Furthermore, in tuber tissues, contact of the plasmalemma with lyase-degraded cell walls alone does not produce membrane damage, but damage does occur if the protoplasts are subjected to osmotic stress. It has, therefore, been suggested that cell injury in enzyme treated tissues is due to some kind of loss in the ability of the degraded cell wall to support the plasmalemma or to protect the plasmodesmata. The nature of this support or protection is not known.

Where photosynthetic tissues are invaded cell death obviously leads to a loss in photosynthetic area. Although host tissues around lesions may cease to export their assimilates there is no evidence that there is increased translocation of metabolites into these areas from other parts of the host.

2. Catabolite Repression and Necrotrophic Activity

Successful necrotrophy obviously depends to a great degree upon copious production by the fungus of extracellular degradative enzymes. Equally obviously as these act upon the various components of host tissue the immediate environment of the necrotroph becomes enriched with degradation products which may affect the physiology of the fungus. In particular, quantitative or qualitative changes, often quite small, in the carbohydrate content of invaded host tissues may markedly affect degradative enzyme synthesis (Albersheim, Jones and English, 1969). There is abundant evidence that cellulase and pectolytic enzyme synthesis are repressed by small amounts of glucose and galactose, or when the inducers of these enzymes are present in slight excess of the requirements for fungal growth (Horton and Keen, 1966a; Keen and Horton 1966; Patil and Dimond, 1968; Biehn and Dimond, 1971; Cooper and Wood, 1975). This must also occur, of course, in obligate saprotrophs and in necrotrophs which are growing saprotrophically in axenic culture (Nisizawa, Suzuki and Nisizawa, 1972). However, in the natural situation, catabolite repression of enzyme

134

synthesis may have far reaching effects on the progress of the necrotroph through host tissues, and a naturally or artificially high carbohydrate status of these may restrict necrotrophic activity, although this seems not to be universally true (Horton and Keen, 1966*b*; Gibbs and Wilcoxson, (1972).

Bibliography

Albersheim, P., Jones, T. M. and English, P. D. 1969. *A. Rev. Phytopath.* 7: 171–194.
Basham, H. G. and Bateman, D. F. 1975. *Physiol. Pl. Path.* 5: 249–262.
Biehn, W. L. and Dimond, A. E. 1971. *Phytopathology.* 61: 242–243.
Calonge, F. O., Fielding, A. H., Byrde, R. J. W. and Akinrefon, O. A. 1969. *J. exp. Bot.* 20: 350–357.
Cooper, R. M. and Wood, R. K. S. 1975. *Physiol. Pl. Path.* 5: 135–156.
Garrett, S. D. 1970. *Pathogenic Root-Infecting Fungi.* Cambridge University Press.
Gibbs, A. F. and Wilcoxson, R. D. 1972. *Physiol. Pl. Path.* 2: 279–288.
Hall, J. A. and Wood, R. K. S. 1970. *Nature, Lond.* 227: 1266–1267.
Hall, J. A. and Wood, R. K. S. 1973. In R. J. W. Byrde and C. V. Cutting (eds.), *Fungal Pathogenicity and the Plant's Response,* p. 19–31, Academic Press: London.
Hall, J. A; and Wood, R. K. S. 1974. *Ann. Bot.* 38: 129–140.
Harley, J. L. 1969. *The Biology of Mycorrhiza.* Hill: London.
Horton, J. C. and Keen, N. T. 1966*a*. *Can. J. Microbiol.* 12: 209–220.
Horton, J. C. and Keen, N. T. 1966*b*. *Phytopathology.* 56: 908–916.
Keen, N. T. and Horton, J. C. 1966. *Can. J. Microbiol.* 12: 443–453.
Lai, M., Weinhold, A. R. and Hancock, J. G. 1968. *Phytopathology.* 58: 240–245.
Lewis, D. H. 1973. *Biol. Rev.* 48: 261–278.
Mount, M. S., Bateman, D. F. and Grant-Basham, H. 1970. *Phytopathology.* 60: 924–31.
Nisizawa, T., Suzuki, H. and Nisizawa, K. 1972. *J. Biochem., Tokyo.* 71: 999–1008.
Patil, S. S. and Dimond, A. E. 1968. *Phytopathology.* 58: 676–82.
Strobel, G. A. 1974. *A. Rev. Pl. Physiol.* 25: 541–566.
Thatcher, F. S. 1939. *Am. J. Bot.* 26: 449–458.
Wheeler, H. and Hanchey, P. 1968. *A. Rev. Phytopath.* 6: 331–350.
Wood, R. K. S., Ballio, A. and Graniti, A. (eds.), 1972. *Phytotoxins in Plant Diseases.* Academic Press: London and New York.

Chapter 9

Biotrophs

Antagonistic biotrophs are widespread and are distributed among taxonomic groups of fungi that are not phylogenetically closely related (Table 7). The majority of antagonistic symbionts of Group 2 (see Chapter 1) are biotrophs and, while some of them may have a capacity for a free-living existence in nature (hemibiotrophs — see Chapter 10) most have no such ability. It is these kinds of fungi that are to be discussed here, although with respect to some aspects of their physiology they have characteristics in common with hemibiotrophs when the latter are in the biotrophic phase of their symbiotic association. It should be pointed out that antagonistic biotrophs also exhibit a number of characteristics which are shown by mutualistic biotrophs that are involved in mycorrhizal and lichen associations.

A major characteristic of antagonistic biotrophic fungi is that they are physiologically extremely specialized, with the concomitant loss in adaptability being reflected in their normally restricted host ranges. Interactions with their hosts, at least when these are higher plants, are distinguished by a number of characteristic features, not all of which may be exhibited by any particular host—fungus combination. First, there is usually, but not invariably, intracellular penetration of the host by the fungus. This is accompanied by nuclear disturbances within both the penetrated and adjacent unpenetrated cells. There is normally minimal tissue damage, although major morphological changes in the host plant may occur. Where photosynthetic tissues are invaded by the biotroph, 'green islands' are maintained around sites of fungal activity while chlorophyll is lost from the surrounding tissues. Finally, in contrast with necrotrophs, biotrophic fungi usually have a capacity to induce the translocation of host metabolites into infected regions. A great deal is known about all these aspects of host-biotroph interactions and a number of reviews and syntheses concerned with them have been published (Shaw, 1963; Thrower, 1965b; Brian, 1967; Yarwood, 1967; Scott, 1972).

The bulk of information available is derived from studies on infections by members of the Uredinales, particularly *Puccinia* species, and by fungi in the Erysiphales, mainly species of *Erysiphe*, on the leaves of cereals. While it is true that much is also known of the effects

Table 7
Taxonomic groups of fungi which contain antagonistic biotrophs

---

Myxomycota
　Plasmodiophoromycetes
　　Plasmodiophorales; Plasmodiophoraceae

Mastigomycotina
　Chytridiomycetes
　　Chytridiales; Chytridiaceae, Synchytriaceae, Cladochytriaceae
　Oomycetes
　　Peronosporales; Peronosporaceae, Pythiaceae, Albuginaceae

Ascomycotina
　Hemiascomycetes
　　Taphrinales; Taphrinaceae
　Plectomycetes
　　Erysiphales; Erysiphaceae
　Pyrenomycetes
　　Hypocreales; Clavicipitaceae,

Basidiomycotina
　Hemibasidiomycetes
　　Uredinales; Melampsoraceae, Pucciniaceae
　　Ustilaginales; Graphiolaceae; Ustilaginaceae; Tilletiaceae
　Hymenomycetes
　　Exobasidiales; Exobasidiaceae

---

of infections by other fungi on leaves, stems, flowers, and subterranean organs, it is inevitable that any broad treatment of host—biotroph physiology must be biased towards a consideration of the rusts and powdery mildews of graminaceous plants. Physiological relationships between many other biotrophic fungi and their hosts show a number of similarities with those involving rusts and powdery mildews, but there is increasing evidence that relationships between the biotrophic 'lower' fungi of the Plasmodiophoromycetes, Chytridiomycetes, and Oomycetes and their hosts may differ in some important respects from those typical of the 'higher' fungi. Finally, it should be emphasized that, while biotrophy is not restricted to fungi which infect higher plants, almost nothing is known of physiological relationships of fungi with cryptogamic hosts, nor of those between aquatic or marine fungi and algae.

1. Features of Intracellular Penetration

Biotrophic fungi invade their hosts in a variety of ways. Those infecting the aerial parts of plants pass into host tissues via stomata, directly through the cuticle, by means of the flower, or through the coleoptile of germinating seeds. Biotrophs of subterranean organs may invade through root hairs or pass directly into roots or tubers through the epidermis. Details of these various means of invasion are well documented and will not be discussed here (Wood, 1967). At an early stage during the successful establishment of a biotroph, a sequence of events occurs during which the normal metabolism of host cells alters as they respond to demands made on them by the fungus. Some of these events may occur soon after host—biotroph contact and before invasion has begun. For example, local hydrolysis of starch occurs in leaves of *Phaseolus* beneath areas where urediospores of the rust fungus *Uromyces phaseoli* are germinating but before penetration has taken place (Schipper and Mirocha, 1969*a*, *b*). The spores of *U. phaseoli*, in common with those of a number of other biotrophic fungi, contain a β-amylase activator of low molecular weight which diffuses through the cuticle and epidermal cells into the mesophyll. The host's hydrolytic enzymes are activated and starch is mobilized, the soluble products presumably being subsequently utilized by the biotroph after successful penetration. Other fungal products may move into the host during prepenetration and it has been shown that urediospores of *Puccinia graminis tritici* lose carbon compounds to the environment during germination which then accumulate in the outer epidermal cell walls. (Daly, Knoche and Wiese, 1967; Ehrlich and Ehrlich, 1970). There is, therefore, reasonable evidence that substances can pass from germinating rust spores into the host and that some of these may in some way modify the infection site, either physically or biochemically, so that further development of the biotroph is favoured.

Movement of metabolites from host to fungus also occurs. For instance, the downy mildew *Bremia lactucae* accumulates ^{3}H-labelled compounds from lettuce cotyledons before penetration, and it has been shown that glucose and leucine are exuded into water droplets lying on the cotyledon's surface (Andrews, 1975).

After invasion there is commonly penetration of living host cells by either the whole thallus of the biotroph, in the case of the Plasmodiophoromycetes and Chytridiomycetes, or, in the case of filamentous fungi, by lateral branches arising from an intercellular mycelium. Intracellular penetration does not invariably take place and in many successful biotrophs, for example some species in the Peronosporales, it may be infrequent, while in others, in particular species in the Taphrinales, it may never occur. There is, therefore, great variation in the nature of the physical contact between biotrophs and their hosts and the nature of these physical relationships has been categorized in detail in relation to their possible physiological signif- icance (Bracker and Littlefield, 1973). It is at these interfaces that host—biotroph contact is at its most intimate and it may be here that many symbiont-directed changes in host physiology originate and are controlled. It should be stressed that, as well as there being diversity in

138

structure, there is also wide variability in the stability of the interfaces. Consequently, they may be either ephemeral or relatively long-lived depending on the degree of compatibility between the fungus and its host.

Intracellular Thalli

Species in the Chytridiomycetes and Plasmodiophoromycetes whose complete thalli occupy the cells of their hosts are by no means uncommon, yet interface structure has been studied in only a few cases. The form of these fungi is much reduced and the nature of the interface with the host appears to be correspondingly simple. Detailed invest-igations have for the most part been confined to the Chytridiomycete *Olpidium brassicae*, which infects the epidermal cells of brassica and lettuce roots, and the Plasmodiophoromycete *Plasmodiophora brassicae* which infects both the hair cells and cortical cells of brassica roots.

During penetration of an epidermal cell by *O. brassicae* the host's plasma membrane is breached so that the young thallus lies within host cytoplasm, being separated from the latter only by its own plasma membrane (Temmink and Campbell, 1969). As the thallus grows no specific associations between it and host cell organelles are established, although small vacuoles may come to surround it so closely that their tonoplasts are almost in contact with the plasma membrane of the thallus. This may or may not have functional significance (Lesemann and Fuchs, 1970). In contrast, in *P. brassicae* there is no penetration of the host plasmalemma in either infected root hair cells or in cortical cells (Williams and McNabola, 1967; Williams and Yukawa, 1967; Aist and Williams, 1971). In both cases the thallus is enclosed in a complex envelope comprising seven layers. This plasmodial envelope is thought to consist of two intimately associated membranes, each being three-layered. The inner one is the plasmodial plasma membrane while the outer one is host-derived, possibly through invagination, detach-ment and modification of the plasmalemma at the time of invasion of the host cell. In some cases within root hairs, the outer membrane appears to be continuous with the tonoplasts of flattened vacuoles that lie between the plasmodial envelope and the host's cytoplasm. The plasmodial envelope is extremely stable and can be separated intact from infected host cells (Keen, Reddy and Williams, 1969; Williams and McNabola, 1970). As the plasmodium grows so does the compound membrane surrounding it and it is possible that biosynthesis of those layers derived from the host's plasmalemma may be controlled by the biotroph. The effect of occupation by *P. brassicae* on the host cell differs from that of *O. brassicae* in one other important respect in that plasmodia of *P. brassicae* stimulate enrichment of the host's cytoplasm which comes to contain abundant ribosomes, dictyosomes, mitochondria and amyloplasts. No such effects are found in adjacent, unoccupied cells even though these are connected to infected cells by numerous plasmodesmata.

Where the whole thallus of a biotroph is immersed in the cytoplasm of its host the fungus is presumably able to absorb major nutrients and other compounds necessary for its growth and reproduction over the whole of the thallus surface. At the same time, fungal metabolites involved in altering the structure and physiology of the infected cell must be free to diffuse out into host cytoplasm. Additional mechanisms of nutrient uptake do, however, exist and at the end of the vegetative phase in plasmodia of *P. brassicae* the outer membrane separates from the inner membrane and the plasmodium may then ingest portions of host cytoplasm through phagocytosis (Williams and McNabola, 1967). Phagocytosis also occurs in *Aphelidium*, a fungus-like organism of dubious status, which infects the alga *Scenedesmus* (Schnepf, Hegewald and Soeder, 1971; Schnepf, 1972).

Haustoria

In a large number of antagonistic biotrophs the only parts of them that are intracellular are haustoria which are intruded through the host cell wall and into its cytoplasm from intercellular hyphae, extracellular hyphae, or from simple extracellular thalli. Haustoria are essentially lateral hyphal branches or somatic appendages of determinate growth, that serve to bring the biotroph into intimate contact with the cytoplasm of the host (Bushnell, 1972). In successful symbioses haustoria coexist with the surrounding host cytoplasm, there being a compatible relationship of varying duration, and during this period haustoria probably facilitate the interchange of substances between the two organisms. A great deal is now known of the structure of haustoria but, although ultrastructural evidence from studies on the haustorium—host interface frequently indicates that they have a specialized function, there is a paucity of direct physiological evidence as to exactly what this function might be. There are two obvious possibilities. They may act as organs for the absorption of major nutrients from the host cell, or they may be involved in more subtle biochemical exchanges which lead to specific modifications in host metabolism that are favourable to the biotroph. These two activities need not, of course, be mutually exclusive.

It has already been pointed out that successful antagonistic biotrophy in many fungi is achieved in the absence of, or with infrequent, intracellular penetration. In relation to this it has been suggested that rust fungi were originally entirely subcuticular or intercellular, that they then evolved primitive spherical or vesicular haustoria and that these, in turn, gradually became larger and more elaborate in shape.

Some rust fungi have reached a state in which they are entirely intracellular (Savile, 1971; Rajendren, 1972a). Therefore, not only do many details of haustorial function remain largely unknown but, in addition, there remains the enigma as to what physiological advantages, if any, are conferred on those fungi which form haustoria in comparison with those that do not.

During the establishment of a haustorium within a host cell a compound structure is formed comprising the hyphal branch itself, components contributed by the host cell and components the origin of which is a matter of debate. The total is commonly referred to as the haustorial apparatus. There is at present no generally accepted terminology for the various structures that make up the haustorial apparatus, different names being used for the same components or identical names for very different structures (Table 8). This has resulted in an unfortunate confusion which is unlikely to be rapidly resolved. The terminology used here closely follows that of Bushnell (1972) which has the virtue of unequivocal simplicity. Details of host cell penetration from extracellular hyphae and the subsequent development of the haustorial apparatus vary depending on the host—symbiont combination involved (Ehrlich and Ehrlich, 1971; Bracker and Littelfield, 1973). Discussion here will be restricted to a consideration of the nature of mature haustoria and, in particular, those of rust fungi. These are by far the most intensively studied but, because of the relative neglect of haustoria of other kinds of fungi, it is not certain as to how far the structure of rust haustoria may be viewed as being the most common type that occurs.

The bulk of the haustorial apparatus is made up of an expanded, nucleate, haustorial body (Figure 52). This arises from a narrow, haustorial neck which is connected through the cell wall of the host to an intercellular haustorial mother cell. The basal portion of the neck is surrounded by a papilla of host material deposited on the host cell wall outside the host protoplast during or after penetration. A sheath, apparently an extension of the papilla, may also be present forming a collar around the haustorial neck. Papilla and sheath seem to be composed principally of callose and both contain irregular masses of what appears to be membraneous material. The apical portion of the haustorial neck and the whole of the haustorial body are enclosed in an extrahaustorial matrix which, it is suggested, contains material secreted by either host or biotroph or both. The haustorial neck and the extrahaustorial matrix are bounded by an extrahaustorial membrane that appears to be continuous with the host's plasmalemma. The extrahaustorial membrane does, however, differ from the host plasmalemma in that the former is non-granular and thicker than the latter. Within the host cytoplasm immediately around the extrahaustorial membrane there is a marked increase in the amount of endoplasmic

Table 8

Various terms currently used to refer to structural components of the haustorial apparatus of antagonistic biotrophic fungi

| | Authority | |
Ehrlich and Ehrlich, 1971	Bushnell, 1972	Bracker and Littlefield, 1973
[a]Encapsulation (also referred to as a sheath in the Erysiphales)	[a]Extrahaustorial matrix	[a]Haustorial sheath or extrahaustorial matrix
Encapsulation boundary or extrahaustorial membrane	[b]Extrahaustorial membrane	Invaginated host plasma membrane or extrahaustorial membrane
Papilla	Papilla	Papilla
Sheath (sometimes also used to refer to the encapsulation)	Sheath or encasement	Collar or apposition or encasement.
Haustorium mother cell	Haustorium mother cell or appressorium	Haustorium mother cell or appressorium

[a]Also sometimes referred to as the zone of apposition in some phycomycetous fungi (Peyton and Bowen, 1963; Chou, 1970).
[b]The sheath membrane of some authors (see Bracker and Littlefield, 1973).

142

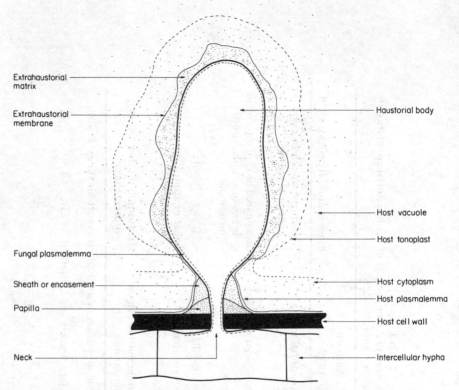

Extrahaustorial matrix

Extrahaustorial membrane

Haustorial body

Host vacuole

Host tonoplast

Fungal plasmalemma

Sheath or encasement

Host cytoplasm

Host plasmalemma

Papilla

Host cell wall

Neck

Intercellular hypha

Figure 52 Diagram of the haustorial apparatus of a biotroph within a host cell.
Not all features illustrated are present in all haustoria

reticulum, in the number and activity of Golgi bodies, and in the
number of mitochondria. Strands of endoplasmic reticulum sometimes
appear to be continuous with the membrane, and vesicles may be
present, some of which are also continuous with the membrane
(Figure 53). The attached vesicles are thought to originate from host
Golgi bodies and appear to move from the host cytoplasm to the
extrahaustorial membrane through which they then discharge substances into the extrahaustorial matrix. Alternatively, they may, of
course, originate at the extrahaustorial membrane and be moving into
the host cytoplasm, carrying substances of fungal origin. Two-way
movement of these vesicles is not impossible.

Haustorial structures in non-rust fungi differ in some respects from
those of rust fungi, but whether these differences are of sufficient
magnitude or significance to merit attempts at a structural classification
of fungal haustoria is not yet certain (Ehrlich and Ehrlich, 1971). The
largest differences occur in the Peronosporales. For example, in host

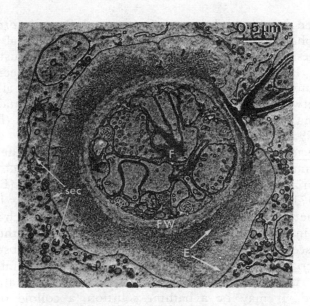

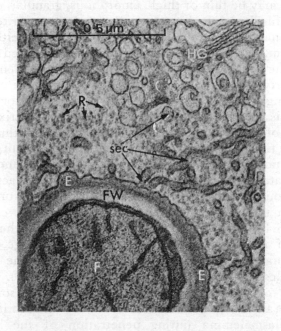

Figure 53 *Albugo candida* Electronmicrographs of an haustorium within radish leaf tissue F, haustorial body; FW, fungal cell wall; E, extra-haustorial matrix; sec, secretory vesicles within host cytoplasm; R, host ribosomes; HG, host Golgi apparatus. From Berlin and Bowen, 1964; by permission of *American Journal of Botany*

cells invaded by *Phytophthora infestans* and *Peronospora parasitica* there is enhanced Golgi activity around the haustorial body, but secretory vesicles and connections between the extrahaustorial membrane and host endoplasmic reticulum have not been observed (Hanchey and Wheeler, 1971). In *P. parasitica* there may also be direct contact between the wall of the haustorial body and the host plasmalemma, the extrahaustorial matrix being absent (Peyton and Bowen, 1963; Chou, 1970).

Any speculation concerning the possible role of the haustorial apparatus must revolve around the nature and function of the extrahaustorial matrix and its bounding membrane (Ehrlich and Ehrlich, 1971; Bushnell, 1972; Bracker and Littlefield, 1973). This region is the interface between fungus and host at which they are in their most intimate association and, if the function of the haustorial body is to secrete substances into the host and receive substances from it, all materials must pass through the matrix and membrane. With respect to the matrix its origin and biochemical nature remain unelucidated. It may be a bathing solution, a colloid or a gel. In appearance it may be thin or thick, amorphous, granular, or layered and may contain fibrils, particulate material, or what appear to be remnants of secretory membranes. In composition it resembles neither fungal nor host cytoplasm. Although its origin cannot be attributed clearly to one or other of the organisms, it is possible that a consequence of the very close proximity of protoplasts of two such dissimilar organisms as a biotrophic fungus and its host could be a mutual secretion of some sort, and that this might result in extrahaustorial matrix production. Maintenance of the matrix would then depend on continued host—biotroph interaction (Zimmer, 1970; Heath and Heath, 1971). The possible functions of the matrix are highly speculative and it remains to be demonstrated whether or not it is biologically active. Labelled carbon compounds have been shown to move from rust mycelium into host cells but there is no accumulation of these in the extrahaustorial matrix (Ehrlich and Ehrlich, 1970). The matrix therefore allows movement of material but does not act as a storage pool for this transported material. It may even be a barrier to the movement of macromolecules from host to biotroph (Manocha, 1975).

The origin of the extrahaustorial membrane is reasonably certain, and it is generally accepted that it is produced through invagination of the host's plasmalemma during penetration of the host cell and expansion of the haustorial body. The membrane cannot be formed by a simple stretching of the plasmalemma, but synthesis of new membrane material must occur as the haustorial body grows. This new membrane material is demonstrably physically different from that of the original plasmalemma and this is taken to mean that it has a specialized function. This view is reinforced by the observed association of numerous host cell components with the membrane (Hardwick, Greenwood and Wood, 1971; Heath, 1972; Littlefield and Bracker, 1972). The most frequently observed association is between host endoplasmic reticulum and the extrahaustorial membrane. This is particularly common in rusts but may be less so in infections by other fungi. The inference is that this close proximity or contact may facilitate transfer between the endoplasmic reticulum and the extra-haustorial membrane (Bracker and Littlefield, 1973). Mention has already been made of the appearance of cytoplasmic vesicles close to the extrahaustorial membrane and an increase in the numbers of Golgi bodies (Peyton and Bowen, 1963; Berlin and Bowen, 1964; Shaw and Manocha, 1965; Van Dyke and Hooker, 1969). There is some indication that the former are secretory bodies of some kind which either play a part in exchange of materials between symbiont and host or are concerned with the development of the extrahaustorial mem-brane, the extrahaustorial matrix or both. Increased activity of the host's Golgi apparatus also strongly suggests that the secretory mechanisms of the host are being stimulated. Which secretions are influenced most and what their role might be is not known.

Despite the dearth of information as to details of physiology of the haustorial apparatus, it is not unreasonable to suppose that such specialized organs must have a specialized and unique function, and that they are primarily involved in some kind of interchange between host and symbiont. There is a great deal of evidence which indicates that substances move from host cells to haustorial biotrophs in large quantities but there is, unfortunately, little evidence that haustoria are bridges for this interchange, either as a whole or in part (Bushnell, 1972; see later sections here). Direct movement of metabolites through host cell walls to and from an intercellular or extracellular mycelium could be the major route for such interchanges in haustorial biotrophs, just as they must be in biotrophs which entirely lack, or infrequently form, haustoria. What evidence there is for a nutritional function comes from studies on *Erysiphe graminis* (Ellingboe, 1968; Mount and Ellingboe, 1969; Slesinski and Ellingboe, 1971). During infection of wheat leaves by this biotroph, growth of extracellular hyphae is not initiated until the primary haustorium arising from the infecting spore is partly formed within its host cell. In addition, labelled compounds supplied to the host move from host to fungus only after the primary haustorium has reached a particular stage of development and has begun to branch. Both these characteristics suggest that the extra-cellular mycelium, at least in its early stages of development, is dependent on functional haustoria for a supply of nutrients. Indirect evidence from protein synthesis studies (to be discussed more fully later) indicates that some haustorial biotrophs may have a limited ability to produce their own proteins, and that in these cases, during establishment of the symbiosis, early haustorium formation may be essential to alleviate this deficiency.

Growth of subcuticular or intercellular hyphae of biotrophs may be accompanied by some enzymatic degradation of the outermost layers of host cell walls, and degradative enzymes are also involved in the penetration of host cell walls during the initiation of haustoria (Ehrlich and Ehrlich, 1963; Calonge, 1969; Edwards and Allen, 1970; Sargent, Tommerup and Ingram, 1973). Production of cutinases and cellulases has been demonstrated in *Erysiphe graminis*, and of cellulases, hemicellulases and pectinolytic enzymes by germinating urediospores of *Puccinia graminis* (Van Sumere, Van Sumere-De Preter and Ledingham, 1957; Kunoh and Akai, 1969; McKeen, Smith and Bhattacharya, 1969). During infection and development within the host, the action of these enzymes is extremely confined, so that there is not the indiscriminate cell wall destruction which is typical of necrotrophs but there is instead localized degradation. This appears to be due to a genetic loss in the ability of many biotrophs to produce the enzymes responsible for host cell collapse, rather than the result of catabolite repression of these enzymes by the sugar-rich environment of host tissue.

2. Effects on Synthesis and Respiration

When host tissues are invaded by a biotroph there is usually an increase in their oxygen consumption, an increase in their RNA and DNA levels and, where green tissues are involved, changes in chlorophyll content (Shaw, 1963; Thrower, 1965b; Heitefuss, 1966). These events are interrelated and are also connected with other alterations in host physiology, for example changes in nutrient distribution and hormone levels.

Changes in Nucleic Acid Metabolism

Much of what is known of nucleic acid metabolism during host—biotroph interaction is derived from investigations on infections involving rust fungi. The nucleoli of germinating urediospores may be reduced in size or be altogether lacking and it has been suggested that this reflects a deficiency in the capacity of the spores for net synthesis of protein during germination, leading to an absolute dependence of the rust on the host plant (Manocha and Shaw, 1967; Mitchell and Shaw, 1969; Manocha and Wisdom, 1971). This is, however, to be doubted since nucleoli have been demonstrated in germ tubes of urediospores and net protein synthesis has been detected during urediospore germination

(Dunkle, Wergin and Allen, 1970; Trocha and Daly, 1970). This synthesis is at the expense of endogenous spore reserves, and subsequent successful growth within the host may depend on a reactivation, by the host, of ribosomal RNA synthesis within the young mycelium, or a contribution of some other kind by the host to the protein-synthesizing mechanisms of the biotroph (Mitchell and Shaw, 1969).

Early consequences of infection by rust fungi are an enlargement of host nuclei and nucleoli, an increase in the RNA content of host tissues and an enhancement of the capacity of the host cells to incorporate labelled precursors into RNA and, albeit rather infrequently, into DNA (Staples and Ledbetter, 1960; Rohringer and Heitefuss, 1961; Rohringer, Samborski and Person, 1961; Bhattacharya and Shaw, 1968; Chakravorty and Shaw, 1971). In addition, in young infections, both haustorial mother cells and the haustorial bodies are rich in RNA (Person, 1960; Whitney, Shaw and Naylor, 1962). Such observations suggest a host-mediated induction of RNA synthesis in the biotroph, but might also indicate induced host protein synthesis, the products being then utilized by the biotroph (Thrower, 1965b). The RNA content of host nucleoli in wheat leaves infected by *Puccinia graminis tritici* may increase by up to 60% (Bhattacharya, Shaw and Naylor, 1968). Since the nucleolus is the site of ribosomal RNA, this implies a stimulation of host rRNA synthesis by the biotroph, a suggestion which has been confirmed through investigations of ^{32}P-orthophosphate incorporation into the rRNA of flax cotyledons infected by *Melampsora lini* (Chakravorty and Shaw, 1971).

Studies on changes in nucleic acid content have also been made for infections involving non-rust fungi, principally *Plasmodiophora brassicae* in cabbage roots. Invasion is through the root hairs within which multinucleate plasmodia develop from initially uninucleate amoeboid cells. Subsequently, root and hypocotyl tissues also become infected, and during development of plasmodia within these there is hypertrophy and hyperplasia of host cells leading to massive gall formation. Upon entry into a hair cell the amoeboid cell lacks a nucleolus, but within 4 hours of invasion the nucleolus appears. The fungal cell then begins to grow rapidly and about 17 hours after infection nuclear division occurs, synchronous divisions taking place at 4 hour intervals until differentiation of plasmodial cytoplasm into spores occurs 72—96 hours after the initial infection. The early appearance of the nucleolus in the amoeboid cell may indicate the rapid establishment of a compatible relationship between the biotroph and its host cell. Plasmodial development within a root hair is paralleled by a two-fold increase in the size of the hair cell nucleolus and cessation of normal linear extension of the hair cell wall, so that the cell becomes clavate and branched (Table 9) (Bhattacharya and Williams, 1971; Williams, Aist

Table 9

Development of uninucleate amoeboid cells of *Plasmodiophora brassicae* and their effect on root hairs of cabbage (after Williams, Aist and Bhattacharya, 1973)

Age of infection (h)	Diam. of fungal cell (μm)	% Amoeboid cells with nucleoli	% Hair cells distorted		Diam. hair cell nucleolus (μm)	
			Infected	Uninfected	Infected	Uninfected
2	0.85	0	13	7	1.21	1.19
4	1.33	30	20	10	1.25	1.04
6	1.54	77	73	13	1.43	1.06
9	1.92	97	87	7	1.82	1.11
15	2.43	100	97	10	1.93	1.17
21	2.63	100	90	13	2.13	1.18

Table 10

Effect of *Plasmodiophora brassicae* on nuclear DNA, total protein, histone and non-histone protein of cabbage root hair cells (after Bhattacharya and Williams, 1971)

Cell condition	DNA	Protein	Histone	Non-histone protein
Infected	11.4	23.3	8.8	14.8
Uninfected	11.4	17.3	10.2	7.1

Figures in arbitrary units of fluorescence corrected for background using microfluorometric techniques. Non-histone protein figure obtained by subtracting values for histone from those for total protein.

and Aist, 1971; Williams, Aist and Bhattacharya, 1973). The increase in host nucleolar volume is accompanied by an increase in nuclear RNA and protein but there is no increase in nuclear DNA (Tables 10 and 11). In addition, the increase in nuclear protein is characterized by a decrease in nuclear histone and an increase in non-histone protein (Table 10), a situation similar to that found in rust-infected wheat leaves (Bhattacharya, Naylor and Shaw, 1965). The absence of an increase in DNA could indicate that *Plasmodiophora* lacks those mechanisms necessary for the induction of DNA synthesis in the host cell, or that synthesis is being blocked in some way by the host cell. Increases in nucleolar size, RNA, and total protein, suggest that host ribosome synthesis is stimulated by the biotroph, while the increase in non-histone protein may indicate that the normal transcriptional processes of the host nucleus are being altered by the presence of the fungus.

Studies on hypocotyl tissues of cabbage show that the nucleic acid metabolism of cells infected by plasmodia differs markedly from that

Table 11

Effect of *Plasmodiophora brassicae* on total nucleic acids, DNA and RNA of cabbage root hair cells (after Williams, Aist and Bhattacharya, 1973)

Cell condition	DNA + RNA	DNA	RNA
Infected	38.4	17.0	21.4
Uninfected	28.3	16.3	12.0

Figures in arbitrary units of fluorescence corrected for background using microfluorometric techniques. RNA figures were obtained by subtracting values for DNA from those for combined DNA–RNA.

of root hair cells (Williams, 1966; Williams, Keen, Strandberg and McNabola, 1968). Infected hypocotyl cells at first retain their normal size but divide rapidly. Then, as the plasmodia within them grow, there is hypertrophy of the host cells together with enlargement of their nuclei and nucleoli. The degree of increase in cell and nucleolar size is much greater than that found in hair cells, there being up to a 30-fold increase in nucleolar volume and up to a 10-fold increase in cell volume. These effects are restricted to occupied cells, neighbouring uninfected cells remaining normal in size, their nuclei and nucleoli being apparently undisturbed. Plasmodial growth is, of course, accompanied by division of the plasmodial nuclei, and there is a direct linear relationship between the number of nuclei within the plasmodium and the volume of the occupied host cell, its nuclear volume and its nucleolar volume. There is also a correlation between host nucleolus size and its relative RNA content. This nuclear and nucleolar hypertrophy results in large increases in the RNA, DNA and histone of host cells (Table 12). The DNA of uninfected cells is distributed normally in a single class but that of infected cells falls into five distinct classes, which suggests different states of ploidy, a $2n$ class (that of uninfected cells) and classes of $4n$, $8n$, $16n$ and $32n$ (Figure 54).

In contrast to the situation obtaining in infected root hairs, a major response of infected hypocotyl cells is enhanced synthesis of DNA. While the plasmodium is small, host cell division continues normally. Then, as plasmodial growth progresses, host cell division is impaired but DNA replication continues. Whether or not host DNA synthesis is stimulated by the biotroph is not known but there is repeated duplication of the diploid DNA complement coincident with enlargement of the host nucleus (Williams, 1966). Unlike DNA, host RNA increases in direct linear relationship to the development of the plasmodium. This has been taken to indicate a delicate balance between the two organisms which allows the host cell to respond in a controlled way to the increasing metabolic demands made on it by the fungus, in particular through increased protein synthesis.

Table 12

Effect of *Plasmodiophora brassicae* on the relative RNA, DNA and histone content of infected cabbage hypocotyl cells (after Williams, 1966)

Cell condition	RNA	DNA	Histone
Infected	13.9	25.4	10.3
Uninfected	2.4	11.2	2.5

Values are arbitrary units obtained using cytophotometric techniques on cells stained for RNA, DNA and nuclear histone.

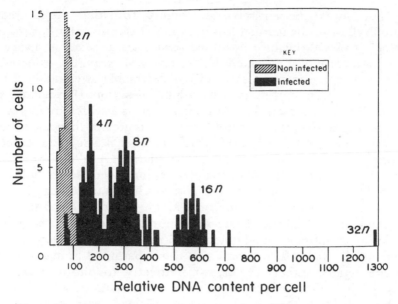

Figure 54   *Plasmodiophora brassicae*, relative DNA content per cell of 65 uninfected and 146 infected cabbage hypocotyl cells taken at random from a transverse section of an infected hypocotyl. From Williams, 1966; by permission of *Phytopathology*

Increases in host DNA may be a characteristic of biotrophic associations which result in pronounced tissue hypertrophy since endopolyploidy has also been found in neoplasms of maize infected by the smut fungus *Ustilago maydis* (Callow and Ling, 1973; Callow, 1975). In contrast, in rust infections where there is no hypertrophy decreases in host DNA have been reported (Whitney, Shaw and Naylor, 1962; Quick and Shaw, 1964).

Photosynthesis and Green Island Formation

During the development of biotrophs on the photosynthetic parts of plants there are usually marked changes in the capacity of the host to synthesize carbon compounds. Transient increases in photosynthesis subsequent to infection have been noted for both powdery mildews and rusts (Allen, 1942; Livne, 1964; Scott and Smillie, 1966). There is, however, a strong possibility that these observed increases are artifacts created by the high concentrations of carbon dioxide used in some investigations (Edwards, 1970). Usually, assimilation of carbon dioxide declines as infection progresses and this is accompanied by a decrease in the chlorophyll content of the affected tissues (Allen, 1942; Last, 1963; Edwards and Allen, 1966; Scott and Smillie, 1966; Doodson, Manners and Myers, 1965). This decrease in photosynthetic activity may not always be entirely due simply to either the decrease in chlorophyll or to the gradual loss in available photosynthetic area as the biotroph grows through or over host tissues. It can occur at a very early stage in infection. For example, in *Erysiphe graminis* infections of barley a decrease in photosynthesis is detectable after only 12 hours, that is at the stage of primary haustorium formation. During later stages of infection, if less than 30% of the leaf area is affected by *E. graminis*, then the actual decrease in photosynthesis is greater than the predicted decrease. Yet if more than 30% of the leaf area is affected the decrease in assimilation is less than the expected value (Last, 1963). Similarly, in wheat infected by *Puccinia striiformis* the decrease in chlorophyll content of leaves occurs both later and to a less marked degree than the decrease in photosynthesis (Doodson, Manners and Myers, 1965).

Within infected tissues chloroplasts normally become reduced in size and suffer membrane breakdown (Whitney, Shaw and Naylor, 1962). As might be expected, this results in the impairment of many processes that are related either directly or indirectly to photosynthesis. For instance, in mildewed barley leaves photoreduction of NADP to $NADPH_2$ declines, there is a loss in the enzymatic activity necessary for both electron transfer and carbon dioxide fixation, and chloroplast ribosomes and polysomes break down (Scott and Smillie, 1966; Dyer and Scott, 1972; Scott, 1972). Although this pattern may prove to be typical there may be exceptions, and chloroplasts isolated from oat leaves infected by *Puccinia coronata* closely resemble physically those from uninfected leaves (Wynn, 1963). Their ability for photophosphorylation is not, apparently, affected. This suggests that the primary reactions in photosynthesis are not drastically altered by this rust infection, but that changes in assimilation are due to effects on the carbon cycle of photosythesis. Loss of chlorophyll from leaves, and the general decrease in assimilation of carbon dioxide, results in general chlorosis and senescence, but within chlorotic areas small regions of apparently healthy tissue may occur. These green islands of retained chlorophyll are located at sites of individual infections or surround zones where the biotroph is sporulating, and within them photosynthesis continues normally (Allen, 1942; Wang, 1961; Thrower, 1965a; Bushnell, 1967; Harding, Williams and McNabola, 1968). They thus provide sites within which synthetic processes are maintained, and within which movement of metabolites from host to biotroph continues, long after the remainder of host tissue has become non-functional (Thrower, 1965b). Enhanced fixation of carbon dioxide in the dark also occurs within green islands induced by rusts, powdery mildews and some of the Peronosporales, but this enhancement seems to be due to the ability of the biotroph itself to dark fix, particularly during sporulation, rather than to an alteration in the metabolism of adjacent host tissues (Mirocha and Zaki, 1965; Daly and Livne, 1966; Mirocha and Rick, 1967; Thornton and Cooke, 1970). The malic enzyme involved in dark fixation has been found in rust urediospores, but it should be noted that both saprotrophic and necrotrophic fungi are capable of this process and that it is not unique to biotrophs. Its significance in biotrophic symbioses is not clear.

Respiration

An almost constant feature of infections by obligate biotrophs is a two- to four-fold increase in the respiration rate of host tissues. It has been stated that the contribution of fungal respiration to this increase is comparatively small and that it is host tissues, under the influence of the biotroph, which contribute mainly to the increase (Allen and Goddard, 1938; Millerd and Scott, 1962). This view has been strongly contested and the alternative view proposed is that the fungus is the major contributor to the rise in respiration (Bushnell and Allen, 1962; Daly, 1967). Whatever the cause, enhanced host respiration does occur and this could be due to either a stimulation of already existing metabolic processes in the host, or to the appearance of new pathways. Available evidence suggests that, in rust and powdery mildew infections, the respiration pattern of host tissues is shifted from a system that is predominantly channelled through glycolysis and the Krebs cycle to one which is dependent on the pentose phosphate pathway (Daly, Sayre and Pazur, 1957; Shaw and Samborski, 1957; Lunderstädt, Heitefuss and Fuchs, 1962; Scott and Smillie, 1963; Scott, 1965). Increases in the activity of the Krebs cycle have also been reported and it is still not clear as to what the relative contribution of this and the pentose phosphate pathway to increased respiration might be (Daly, 1967). In addition, the Peronosporales may differ from rusts and powdery mildews in their effects on host respiration. For example, when radish cotyledons are infected with *Albugo candida* respiration increases two- to three-fold, but this is due to a stimulation of already existing pathways and there is no involvement of the pentose phosphate pathway (Williams and Pound, 1964).

Much debate centres around the significance of changes in photosynthesis and respiration and on the possible interrelationships between them (Thrower, 1965b; Scott, 1972). An increase in the activity of the pentose phosphate pathway results in increased amounts of $NADPH_2$. This could participate in the enhancement of synthetic processes, particularly lipid and sugar alcohol synthesis, which are important to the metabolism of rusts and powdery mildews. The pentose phosphate pathway might also provide an increased supply of pentoses for the synthesis of nucleic acids, and ultimately for protein synthesis.

In etiolated barley leaves there is no increase in respiration upon infection by *Erysipbe graminis* but there is up to a four-fold increase in green leaves of the same age. The onset of respiration increase occurs simultaneously with a decrease in photosynthesis in an area limited to chloroplast destruction beneath the developing biotroph (Scott and Smillie, 1963, 1966). Since both the pentose phosphate pathway and photosynthesis are major sources of $NADPH_2$, the increase in repiration may be linked to, and in some way compensate for, a progressive loss in the ability of host cells to photoreduce NADP to $NADPH_2$.

3. Changes in Nutrient Distribution

Many kinds of metabolites accumulate within and around sites of biotrophic infections (Yarwood and Jacobson, 1955). Of these metabolites it is carbon compounds which have been studied in most detail, and it is generally accepted that successful development of biotrophs within their hosts requires an adequate supply of carbohydrate from host tissues (Scott, 1972; Lewis, 1974). Most, if not all, biotrophs have a capacity to influence the distribution of carbohydrates within host tissues so that a typical feature of infections involving rust fungi, powdery mildews, *Albugo* species, and *Plasmodiophora brassicae*, is the accumulation of one or more host carbohydrates at infection sites (Bushnell and Allen, 1962; Williams and Pound, 1964; Williams, Keen, Strandberg and McNabola, 1968; Keen and Williams, 1969; Holligan, Chen and Lewis, 1973; Long and Cooke, 1974). Glucose, fructose, and sucrose are commonly involved but there is also large-scale deposition of starch or other storage polysaccharides prior to sporulation of the fungus. These then disappear as spore production progresses (Allen, 1942; Wang, 1961; Thrower, 1965a; Mirocha and Zaki, 1966; Bushnell, 1967). In leaves, zones of starch deposition in young infections may correspond in location and extent to the green islands which subsequently develop as host tissue senesces. Accumulation of soluble and insoluble carbohydrates is often accompanied by an increase in weight of the infected organ. Polysaccharide accumulation does not normally occur in uninfected senescing leaves.

In the photosynthetic parts of host plants the site of starch deposition is in those chloroplasts adjacent to fungal hyphae, while in non-photosynthetic tissues starch grains are found within the cytoplasm of host cells (Keen and Williams, 1969; MacDonald and Strobel, 1970). The mechanisms underlying increased starch formation are largely unknown but there is some evidence for direct intervention of the biotroph in the host's starch synthesizing systems. In healthy wheat leaves the rate of starch synthesis is determined by the activity of ADP-glucose phosphorylase, an enzyme requiring activation by 3-phosphoglyceric acid, but which is inhibited by the presence of orthophosphate. In wheat leaves infected with *Puccinia striiformis*, changes in starch synthesis can largely be accounted for by changes in orthophosphate levels at infection sites (MacDonald and Strobel, 1970). It has been observed that urediospores of *Puccinia graminis tritici* are rich in inorganic polyphosphates, molecules that have a role in regulating cellular orthophosphate levels in fungi, and the metabolism of these polyphosphates may control orthophosphate levels not only in rust mycelium, but also in surrounding host cells. Sequestration of orthophosphate as polyphosphate during inorganic polyphosphate synthesis might lead to an increase in ADP-glucose pyrophosphorylase activity and, subsequently, enhanced starch synthesis (Bennett and Scott, 1971; Scott, 1972). Large increases in the amount of starch and reducing sugars at infection sites are also connected with increased activity of acid invertase. After hydrolysis of sucrose to glucose and fructose the latter two sugars are then taken up and metabolized by the biotroph but, in addition, where there is a surplus of soluble carbohydrate, particularly sucrose, hydrolysis provides hexoses for starch synthesis within the chloroplast (Figure 55). When a host normally stores insoluble fructans rather than insoluble glucans the activity of invertase similarly leads to fructan synthesis and deposition of these compounds within the vacuoles of host cells (Figure 56) (Long, Fung, McGee, Cooke and Lewis, 1975). Invertase thus mediates a system by which the excess soluble carbohydrate occurring at infection sites is converted to polysaccharides that are less osmotically active. Potential osmotic interference with the physiology of the biotroph is in this way prevented.

The accumulation of carbohydrates and other metabolites within infected tissues is contributed to by two interlinked processes; nutrient mobilization with short distance movement of host metabolites towards the biotroph, and long distance translocation to infection sites from the rest of the plant. In general, when a leaf is infected, there is a decrease in the amount of photosynthate exported from the leaf, coupled with an increase in the amount of assimilate imported by it, particularly if infection is severe. This situation contrasts with that obtaining in infections by necrotrophic fungi where, although a decrease in export from infected tissues has been observed, increased import does not seem to normally occur (Livne and Daly, 1966; Thrower and Thrower, 1966; Holligan, Chen, McGee and Lewis, 1974). Translocation of photosynthates to roots may be markedly diminished by general infections of the leaves, with a consequent reduction in root growth (Siddiqui and Manners, 1971).

The essentially one way movement of carbohydrates from the host to sites of biotrophic activity may be maintained in a variety of ways. Accumulation at sites of rust and powdery mildew infections is inhibited under anaerobic conditions and is sensitive to respiratory

156

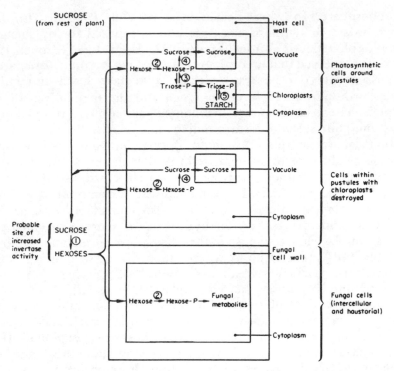

Figure 55 Schematic representation of probable site and pathway of starch accumulation in leaves infected by a biotroph. Enzymes and enzyme sequences involved: 1, invertase; 2, hexokinase; 3, glycolysis; 4, sucrose synthetic system; 5, Calvin cycle and starch synthetase. From Long, Fung, McGee, Cooke and Lewis, 1975; by permission of *New Phytologist*

inhibitors (Shaw and Samborski, 1957). This indicates that carbohydrate accumulation may to some degree depend on the increase in respiration which is normally a consequence of infection. Continual removal of carbon compounds through respiration could lead to infection sites acting as metabolic sinks which would then bring about both local accumulation and long distance translocation. This process would, however, seem to be a wasteful one and would not prevent the re-utilization of accumulated carbohydrates by host tissues around infection sites. It is now well established that carbon compounds moving from host to biotroph accumulate within the hyphae in the form of specifically fungal products. In infections by species of Basidiomycotina and Ascomycotina these are usually acyclic polyols, principally arabitol and mannitol, together with trehalose and glycogen (Daly, Inman and Livne, 1962; Livne, 1964; Thrower and Lewis, 1973; Holligan, Chen, McGee and Lewis, 1974; Holligan, McGee and Lewis,

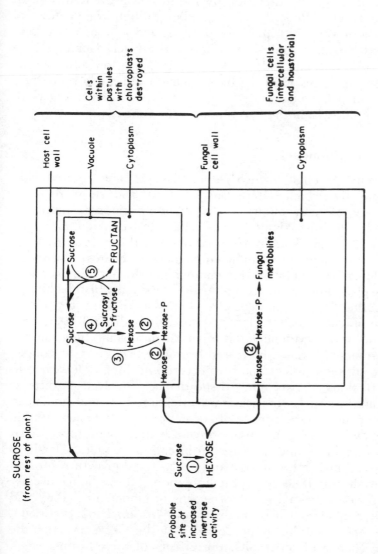

Figure 56 Schematic representation of probable site and pathway of fructan accumulation in leaves infected by a biotroph. Enzymes and enzyme sequences involved: 1, invertase; 2, hexokinase; 3, sucrose synthetic system; 4, sucrose–sucrose 1-fructosyl-transferase; 5, fructan–fructan 1-fructosyl-transferase. From Long, Fung, McGee, Cooke and Lewis, 1975; by permission of *New Phytologist*

158

1974). In common with lichenized and mycorrhizal fungi, antagonistic
biotrophs convert host sugars to fungal carbohydrates, so maintaining
flow of carbon compounds towards sites of fungal activity (see
Chapters 12 and 13). Since these fungal products cannot be metabolized by host cells, their re-utilization by the host is prevented (Smith,
Muscatine and Lewis, 1969). Species of Plasmodiophoromycetes and
Oomycetes do not appear to produce polyols, and although some are
capable of synthesizing trehalose in others it is entirely lacking (Keen
and Williams, 1969; Long and Cooke, 1974; Thornton and Cooke,
1974). In these organisms source—sink movement of host carbohydrate
may be maintained by means of the conversion of soluble sugars to
glycogen or to lipids within the cytoplasm of the fungus.

4. Involvement of Hormones

The mechanisms through which green islands are formed and
maintained and by which alterations in long and short distance
translocation are brought about are probably hormonal. Increased
hormone levels in host tissue are a characteristic of biotrophic
infections but it is not known to what extent this is due to their
production by the fungus or to induced changes in the host's hormone
metabolism. Specific details of the manner in which these increases
affect various physiological processes in host tissue are also lacking, a
situation reflecting the general lack of certainty as to the exact role of
hormones in uninfected higher plants. Despite these difficulties, a great
deal of circumstantial evidence points to an important function for
hormones in infected tissues. Increased hormone levels are, of course,
also a feature of some infections by necrotophic fungi. However, in
biotrophic infections, the minimal tissue damage involved permits at
once both transport of hormones from infection sites and enhanced
nutrient translocation to the sites, a situation not found in necrotrophic
associations (Lewis, 1973). Hormones obviously control morphogenesis
of the host and a hormonal increase on infection frequently leads to
induced abnormal host morphology and exaggerated growth responses
in both biotrophic and necrotrophic infections, these being usually
attributable to supra-optimal levels of auxins or gibberellins (Daly and
Inman, 1958; Sequeira, 1963; Bailiss and Wilson, 1967; Sequeira,
1973). While increased hormone levels may be necessary for the
efficient maintenance of biotrophs, the effects of these hormones on
host growth are probably incidental to this more important role. High
hormone levels around infection sites may be essential to the biotroph
for a number of reasons, but a general change in growth regulators
throughout the plant might not. In this case abnormally growing host
tissues would be a fortuitous, uncontrollable and perhaps even
undesirable, consequence of what is initially a localized enhancement of

hormone synthesis. The production of morphologically abnormal tissues *per se* is thus, perhaps, not an essential prerequisite for successful symbiosis.

Hyperauxiny has been demonstrated in infections by rust and powdery mildew fungi, species of *Taphrina*, and *Plasmodiophora brassicae*, there commonly being high levels of indole-3-acetic acid (IAA) (Daly and Inman, 1958; Shaw and Hawkins, 1958; Ingram, 1969; Kavanagh, Reddy and Williams, 1969; Sziráki, Balázs and Király, 1975). In uninfected hosts localized enhancement of IAA levels, through either auxin production by the biotroph or stimulation of the host's auxin synthesizing systems, could have the effect of directing nutrient flow to infection sites. Auxin production might also enhance synthesis, particularly of RNA and protein, in host tissues surrounding the biotroph, so that there is consequent movement of metabolites into the affected tissues which are acting as a metabolic sink (Thrower, 1965b). Details of the exact origin of high levels of IAA in infected tissues are generally lacking. A number of biotrophs have been shown to be capable of synthesizing IAA in axenic culture, particularly if they are supplied with tryptophane, but the significance of their contribution in the symbiotic situation is not clear (Crady and Wolf, 1959; Sommer, 1961; Srivasta and Shaw, 1962). There is some indirect evidence that in rust infections hyperauxiny may be the result of interference by the biotroph with the host's auxin-degrading enzymes, so producing excess synthesis of host hormones (Sequeira, 1973). A rather different situation that also results in hyperauxiny has been demonstrated in *Brassica* species which are susceptible to *Plasmodiophora brassicae* (Butcher, El-Tigani and Ingram, 1974). Host tissues contain the indole compound glucobrassicin and also an enzyme glucosinolase which can degrade glucobrassicin to indole acetonitrile (IAN) a precursor of IAA. In healthy plants glucobrassicin and glucosinolase are compartmentalized separately but infection by *P. brassicae* interferes with cell integrity in such a way that the two are exposed to one another, hence releasing IAN so that IAA activity subsequently increases. One further effect of excess IAA levels in biotrophic infections may be to directly divert the host's respiratory pathways to the pentose phosphate pathway. However, the evidence for this is conflicting although the possibility is still open (Shaw, Samborski and Oaks, 1958; Sequeira, 1963).

Cytokinins have also been strongly implicated in producing many symptoms which are characteristic of biotrophic infections, since they can induce green island formation in uninfected senescing leaves and can direct long distance phloem transport of nutrients (Thrower, 1965a; Pozsár and Király, 1966; Bushnell, 1967). Compounds with high cytokinin activity have been obtained from urediospores of rusts, conidia of powdery mildews, and axenic cultures of *Taphrina* species, and their presence has been demonstrated in tissues infected by biotrophs (Bushnell and Allen, 1962; Thrower, 1965a; Dekhuijzen and Staples, 1968; Reddy and Williams, 1970; Dekhuijzen and Overeem, 1971; Kern and Naef-Roth, 1975; Sziráki, Balázs and Király 1975). Despite this body of evidence it is still not yet established what the exact significance of these compounds is in the production of disease symptoms. Whether they act alone or synergistically with auxins and gibberellins is not clear, nor has it been convincingly demonstrated that nutrient mobilization and normal photosynthesis occur in cytokinin-induced green islands (Scott, 1972). Furthermore, in some powdery mildew infections it has been shown that both hyphal growth and primary haustorium formation are inhibited by cytokinins (Dekker, 1963; Cole and Fernandez, 1970). The contention that the development of biotrophic infection is favoured by high cytokinin levels cannot at the moment be substantiated (Lewis, 1973).

5. Dual and Axenic Culture

The greater part of our knowledge of the nature of host—biotroph interactions has come from investigations on whole host plants or on detached parts of them. In using such methods there is the obvious attendant problem as to how to correctly interpret the behaviour of a dual entity when, in many circumstances, it is difficult to discern which effects are due to host activity and which to those of the fungus. It has been argued that important advantages would be gained if nutritional, physiological and metabolic studies on antagonistic biotrophs were carried out using more simple systems (Scott and MacLean, 1969; Scott, 1972; Ingram and Tommerup, 1973). At present two possibilities present themselves, involving either the use of infected host tissue cultures or of fungal material entirely free from host tissues. Both systems allow experiments to be carried out in a controlled physical and nutritional environment in the absence of alien micro-organisms. It is now, however, becoming more and more apparent that a number of problems exist which render such methods rather less valuable than was at first supposed.

Several downy mildew and rust fungi have been successfully grown on host callus tissues upon which they sporulate and within which they form apparently normal haustoria (Griffin and Coley-Smith, 1968; Scott and Maclean, 1969; Tiwari and Arya, 1969; Ingram and Joachim, 1971; Ingram and Tommerup, 1973; Harvey, Chakravorty, Shaw and Scrubb, 1974; Robb, Harvey and Shaw, 1975a, b). Although some metabolic studies have been carried out on infected calluses the value of the dual culture system as an approach to the elucidation of aspects of host—biotroph physiology may be seriously questioned. Quite obviously, the metabolism of callus tissue differs very markedly from that of the host from which it was derived, one of the most important differences, with respect to studies on fungi which normally infect leaves or stems, being that they commonly lack chlorophyll. In addition, they usually have high meristematic activity combined with a low degree of differentiation. They may also produce fungitoxic secondary metabolites and may alter genetically through either mutation or an increase in ploidy (Maheshwari, Hildebrandt and Allen, 1967; Ingram and Tommerup, 1973). In one instance normal host—biotroph specificity has been found to be lost, the rust fungus *Cronartium ribicola* infecting callus from the non-host species *Pseudotsuga menziesii* as successfully as it does its host species *Pinus monticola*. Infection of the latter by *C. ribicola* also brings about degenerative changes in host cells, at some distance from the site of activity of the fungus, which are untypical of the interaction of host and biotroph in the natural situation (Harvey and Grasham, 1971; Robb, Harvey and Shaw, 1975*a*, *b*).

There are a number of ways in which it is possible to obtain mycelium of biotrophs free from host tissues, and hyphal material of the rust fungi *Uromyces phaseoli* and *Melampsora lini* has been separated from infected leaves by physical disruption of the latter or through the use of cell wall degrading enzymes (Dekhuizen, Singh and Staples, 1967; Lane and Shaw, 1972). Dual biotroph—callus cultures also lend themselves to manipulation, and calluses have yielded host-free colonies of the downy mildew fungi *Peronospora farinosa* and *Pseudoperonospora humuli*, these being formed on the agar a little distance from host tissue. The continued development of these two species seems to depend on physical connection with the parent callus, and if the connecting hyphae are severed mycelial growth ceases (Griffin and Coley-Smith, 1968; Ingram and Joachim, 1971). True, but limited, axenic growth of two other downy mildew species has been obtained and *Peronospora parasitica* and *Sclerospora graminicola* will grow on semisynthetic media upon which both sporulate (Guttenburg and Schmoller, 1958; Tiwari and Arya, 1969). It is, however, doubtful whether any of these methods can at the moment consistently provide sufficient fungal material in a condition suitable for nutritional and physiological investigations to be made upon it.

In contrast to the relative lack of success in establishing persistent axenic colonies of downy mildews, attempts to grow host-free cultures of rust fungi have met with an increasing number of positive results and species of *Cronartium*, *Gymnosporangium*, *Melampsora*, *Puccinia*, and *Uromyces* have been obtained and maintained in axenic culture. Inoculum has either been in the form of material obtained from callus culture or, more commonly, as urediospores from infected host plants (Harvey and Grasham, 1970, 1974; Wolf, 1974).

The most intensive studies on axenically cultured rusts have been carried out on various races of *Puccinia graminis tritici*. Incubation of urediospores on nutrient media at relatively low temperatures (16—18°C) with a water-saturated atmosphere allows a proportion of the germinating spores to give rise to saprotrophic colonies. Saprotrophic growth is stimulated in those areas on the agar where urediospores are most densely seeded. Germination is erratic and there is a long lag period before axenic growth begins. In addition, while some of the colonies may continue to grow for several months, their growth rate may decline markedly during this period, and they may become pigmented and lose their viability (Williams, Scott and Kuhl, 1966; Wong and Willetts, 1970; Bushnell and Stewart, 1971). Similar behaviour has been observed during axenic growth of *Puccinia coronata avenae* and *Uromyces dianthi* (Jones, 1972, 1974). Persistently growing axenic colonies of *P. graminis tritici* may produce urediospores and also, on ageing, teliospores. The urediospores will successfully infect the leaves of wheat seedlings but only if the epidermis is first removed to expose the underlying mesophyll cells (Williams, Scott, Kuhl and Maclean, 1967; Bushnell, 1968). Urediospores and teliospores are also produced by axenic colonies of *Melampsora lini*, but in this species urediospores can infect intact leaves, and this ability is not lost even after 12 months growth in culture (Turel, 1969, 1971).

Although rust fungi can be grown axenically with, in at least some cases, relative ease, the entity which grows and develops saprotrophically on nutrient agar often differs in some important respects from the biotrophic material from which it originated. In *P. graminis tritici* the urediospores used as inoculum for axenic culture are dikaryotic, as is their parent mycelium, but two kinds of axenic colonies may arise from these spores. First, there are fast growing colonies with binucleate cells and, second, slower growing colonies with uninucleate cells (Maclean and Scott, 1970). The former soon cease to grow, due to either staling or to the production of stromatal tissue, and it is the monokaryotic, slower-developing colonies, which have a less-developed tendency to stale or produce stromata, that in some cases may eventually comprise the majority of persistently growing mycelia (Maclean, 1974). Colonies with uninucleate cells may also develop from dikaryotic isolates originating from urediospores by means of somatic recombination, and there is evidence that in some cases their single nuclei may be diploid. These diploid lines are unable to infect wheat leaves (Maclean, Scott and Tommerup, 1971; Williams and Hartley, 1971; Williams, 1975). Some isolates cannot readily synthesize new nuclear material so that this deficiency is alleviated during axenic growth by continual migration of old nuclei into the younger regions of the colony. The terminal cells of hyphae may at first lack nuclei, then become mono- di- or even tri-karyotic as they age. The cells of the colony centre, after losing their nuclei, eventually degenerate (Rajendren, 1972*b*).

These abnormal situations that arise with some races of *P. graminis tritici* may not occur so commonly in other rust fungi, and in *Melampsora lini*, for example, the cells of the axenic mycelium are persistently binucleate (Coffey, Bose and Shaw, 1970). The reasons for erratic growth, cessation of development, and nuclear changes in axenic culture are still not entirely clear but may obviously be related to the ability of the fungus to make the necessary physiological and metabolic transition from a state of biotrophy to one of saprotrophy. Why at least some successful adaptations to saprotrophy should be accompanied by, or perhaps even depend upon, loss of the normal dikaryotic condition is not known.

When growing axenically rust fungi can utilize a wide range of soluble carbohydrates as sole sources of carbon. They are self-sufficient for vitamins, but lack the ability to utilize inorganic sources of nitrogen and may require growth factors present in yeast extract or peptone (Coffey and Shaw, 1972; Jones, 1973). A major feature of their nutrition in axenic culture is an inability to incorporate inorganic forms of sulphur into amino acids, so that they have a requirement for an exogenous supply of sulphur-containing amino acids. Cysteine, homocysteine or methionine all support growth in *P. graminis tritici*, but in *M. lini* methionine will not substitute for cysteine (Kuhl, Maclean, Scott and Williams, 1971; Coffey and Allen, 1973). Mycelium of *P. graminis tritici* takes up ^{35}S-sulphite but is unable to reduce it. This species will, however, incorporate ^{35}S-sulphide but still has a requirement for a sulphur amino acid (Howes, 1973). Incorporated, labelled sulphide can be detected in culture filtrates in cysteine, S-methylcysteine, glutathione and cysteinylglycine. Appreciable amounts of methionine are found in protein within the mycelium yet little methionine appears in culture filtrates. This implies that the requirement of this species for exogenous sulphur compounds may be caused by continual and excessive loss of cysteine and some of its derivatives from the hyphae (Howes and Scott, 1972). This might also explain why in some strains of *P. graminis tritici*, increased urediospore density increases the chance of the successful establishment of axenic colonies, since in this situation leakage would be at least partially alleviated by reabsorption. Where urediospores are relatively widely separated sulphur amino acids would rapidly diffuse away. With respect to this possibility it has been observed that sometimes application of cysteine aids axenic growth in areas of low spore density.

While the objectives of experimental investigations on dual and axenic cultures of antagonistic biotrophic fungi are laudable ones it is obvious that, at least with rusts at the present time, the amount of

164

information so far obtained from such studies is extremely limited. It is also very difficult to relate the results so far obtained to natural rust infection. Until techniques are devised which produce genetically stable, long-lived axenic cultures of biotrophic fungi this situation is unlikely to be radically changed.

Bibliography

Aist, J. R. and Williams, P. H., 1971. *Can. J. Bot.*, 49: 2023—2034.
Allen, P. J., 1942. *Am. J. Bot.*, 29: 425—435.
Allen, P. J. and Goddard, D. R., 1938. *Am. J. Bot.*, 25: 613—621.
Andrews, J. H., 1975. *Can. J. Bot.*, 53: 1103—1115.
Bailiss, K. W. and Wilson, I. M., 1967. *Ann. Bot.*, 31: 195—211.
Bennett, J. and Scott, K. J., 1971. *Physiol. Pl. Path.*, 1: 185—195.
Berlin, J. D. and Bowen, C. C., 1964. *Am. J. Bot.*, 51: 452—455.
Bhattacharya, P. K., Naylor, J. M. and Shaw, M., 1965. *Science N.Y.*, 150: 1605—1607.
Bhattacharya, P. K. and Shaw, M., 1968. *Can. J. Bot.*, 46: 93—99.
Bhattacharya, P. K., Shaw, M. and Naylor, J. M., 1968. *Can J. Bot.*, 46: 11—16.
Bhattacharya, P. K. and Williams, P. H., 1971. *Physiol. Pl. Path.*, 1: 167—175.
Bracker, C. E. and Littlefield, L. J., 1973. In R. J. W. Byrde and C. V. Cutting (eds.), *Fungal Pathogenicity and the Plant's Response*, p. 159—318, Academic Press, London.
Brian, P. W., 1967. *Proc. R. Soc.*, B, 168: 101—118.
Bushnell, W. R., 1967. In C. J. Mirocha and I. Uritani (eds.), *The Dynamic Role of Molecular Constituents in Plant—Parasite Interactions*, p. 21—39, Bruce, St. Paul, Minnesota.
Bushnell, W. R., 1968. *Phytopathology*, 58: 526—527.
Bushnell, W. R., 1972. *A. Rev. Phytopath.*, 10: 151—176.
Bushnell, W. R. and Allen, P. J., 1962. *Pl. Physiol.*, Lancaster, 37: 751—758.
Bushnell, W. R. and Stewart, D. M., 1971. *Phytopathology*, 61: 376—379.
Butcher, D. N., El-Tigani, S. and Ingram, D. S., 1974. *Physiol. Pl. Path.*, 4: 127—140.
Callow, J. A., 1975. *New Phytol.*, 75: 253—257.
Callow, J. A. and Ling, I. T., 1973. *Physiol. Pl. Path.*, 3: 489—494.
Calonge, F. D., 1969. *Arch. Mikrobiol.*, 67: 209—225.
Chakravorty, A. K. and Shaw, M., 1971. *Biochem. J.*, 123: 551—557.
Chou, C. K., 1970. *Ann. Bot.*, 34: 189—204.
Coffey, M. D. and Allen, P. J., 1973. *Trans. Br. mycol. Soc.*, 60: 245—260.
Coffey, M. D., Bose, A. and Shaw, M., 1970. *Can. J. Bot.*, 48: 773—776.
Coffey, M. D. and Shaw, M., 1972. *Physiol. Pl. Path.*, 2: 37—46.
Cole, J. S. and Fernandez, D. L., 1970. *Ann. appl. Biol.*, 66: 239—243.
Crady, E. E. and Wolf, F. T., 1959. *Physiologia Pl.*, 12: 526—533.
Daly, J. M., 1967. In C. J. Mirocha and I. Uritani (eds.), *The Dynamic Role of Molecular Constituents in Plant—Parasite Interactions*, p. 144—161, Bruce, St. Paul, Minnesota.
Daly, J. M. and Inman, R. E., 1958. *Phytopathology*, 48: 91—97.
Daly, J. M., Inman, R. E. and Livne, A., 1962. *Pl. Physiol.*, Lancaster, 37: 531—538.
Daly, J. M., Knoche, H. W. and Wiese, M. V., 1967. *Pl. Physiol.*, Lancaster, 42: 1633—1642.
Daly, J. M. and Livne, A., 1966. *Phytopathology*, 56: 164—169.
Daly, J. M., Sayre, R. M. and Pazur, J. H., 1957. *Pl. Physiol.*, Lancaster, 32: 44—48.

Dekhuijzen, H. M. and Overeem, J. C., 1971. *Physiol. Pl. Path.*, 1: 151—161.
Dekhuijzen, H. M., Singh, H. and Staples, R. C., 1967. *Contr. Boyce Thompson Inst. Pl. Res.*, 23: 367—372.
Dekhuijzen, H. M. and Staples, R. C., 1968. *Contr. Boyce Thompson Inst. Pl. Res.*, 24: 39—51.
Dekker, J., 1963. *Nature, Lond.*, 197: 1027.
Doodson, J. K., Manners, J. G. and Myers, A., 1965. *J. exp. Bot.*, 16: 304—317.
Dunkle, L. D., Wergin, W. P. and Allen, P. J., 1970. *Can. J. Bot.*, 48: 1693—1695.
Dyer, T. A. and Scott, K. J., 1972. *Nature, Lond.*, 236: 237—238.
Edwards, H. H., 1970. *Pl. Physiol., Lancaster*, 45: 594—597.
Edwards, H. H. and Allen, P. J., 1966. *Pl. Physiol., Lancaster*, 41: 683—688.
Edwards, H. H. and Allen, P. J., 1970. *Phytopathology*, 60: 1504—1509.
Ehrlich, H. G. and Ehrlich, M. A., 1963. *Am. J. Bot.*, 50: 123—130.
Ehrlich, M. A. and Ehrlich, H. G., 1970. *Phytopathology*, 60: 1850—1851.
Ehrlich, M. A. and Ehrlich, H. G., 1971. *A. Rev. Phytopath.*, 9: 155—184.
Ellingboe, A. H., 1968. *A. Rev. Phytopath.*, 6: 317—330.
Griffin, M. J. and Coley-Smith, J. R., 1968. *J. gen. Microbiol.*, 53: 231—236.
Guttenburg, H. von and Schmoller, H., 1958. *Arch. Mikrobiol.*, 30: 268—279.
Hanchey, P. and Wheeler, H., 1971. *Phytopathology*, 61: 33—39.
Harding, H., Williams, P. H. and McNabola, S. S., 1968. *Can. J. Bot.*, 46: 1229—1234.
Hardwick, N. W., Greenwood, A. D. and Wood, R. K. S., 1971. *Can. J. Bot.*, 49: 383—390.
Harvey, A. E., Chakravorty, A. K., Shaw, M. and Scrubb, L. M., 1974. *Physiol. Pl. Path.*, 4: 359—371.
Harvey, A. E. and Grasham, J. L., 1970. *Can. J. Bot.*, 48: 71—73.
Harvey, A. E. and Grasham, J. L., 1971. *Can. J. Bot.*, 49: 881—882.
Harvey, A. E. and Grasham, J. L., 1974. *Phytopathology*, 64: 1028—1035.
Heath, M. C., 1972. *Phytopathology*, 62: 27—38.
Heath, M. C. and Heath, I. B., 1971. *Physiol. Pl. Path.*, 1: 277—287.
Heitefuss, I. R., 1966. *A, Rev. Phytopath.*, 4: 221—244.
Holligan, P. M., Chen, C. and Lewis, D. H., 1973. *New Phytol.*, 72: 947—955.
Holligan, P. M., Chen, C., McGee, E. E. M. and Lewis, D. H., 1974. *New Phytol.*, 73: 881—888.
Holligan, P. M., McGee, E. E. M. and Lewis, D. H., 1974. *New Phytol.*, 73: 873—879.
Howes, N. K., 1973. *J. gen. Microbiol.*, 76: 345—354.
Howes, N. K. and Scott, M. J., 1972. *Can. J. Bot.*, 50: 1165—1170.
Ingram, D. S., 1969. *J. gen. Microbiol.*, 56: 55—57.
Ingram, D. S. and Joachim, I., 1971. *J. gen. Microbiol.*, 69: 211—220.
Ingram, D. S. and Tommerup, I. C., 1973. In R. J. W. Byrde and C. V. Cutting (eds.), *Fungal Pathogenicity and the Plant's Response*, p. 121—137, Academic Press, London.
Jones, D. R., 1972. *Trans. Br. mycol. Soc.*, 58: 29—36.
Jones, D. R., 1973. *Physiol. Pl. Path.*, 3: 379—386.
Jones, D. R., 1974. *Trans. Br. mycol. Soc.*, 63: 593—594.
Kavanagh, J. A., Reddy, M. N. and Williams, P. H., 1969. *Phytopathology*, 59: 1035.
Keen, N. T., Reddy, M. N. and Williams, P. H., 1969. *Phytopathology*, 59: 637—644.
Keen, N. T. and Williams, P. H., 1969. *Pl. Physiol., Lancaster*, 44: 748—754.
Kern, H. and Naef-Roth, S., 1975. *Phytopath. Z.*, 83: 193—222.
Kuhl, J. L., Maclean, D. J., Scott, K. J. and Williams, P. G., 1971. *Can. J. Bot.*, 49: 201—209.

166

Kunoh, H. and Akai, S., 1969. *Mycopath. Mycol. appl.*, 37: 113—118.
Lane, W. D. and Shaw, M., 1972. *Can. J. Bot.*, 50: 2601—2604.
Last, F. T., 1963. *Ann. Bot.*, 27: 685—690.
Lesemann, D. E. and Fuchs, W. H., 1970. *Arch. Mikrobiol.*, 71: 20—30.
Lewis, D. H., 1973. *Biol. Rev.*, 48: 261—278.
Lewis, D. H., 1974. In M. J. Carlile and J. J. Skehel (eds.), *Evolution in the Microbial World*, p. 367—392. 24th *Symp. Soc. gen. Microbiol.*, Cambridge University Press.
Littlefield, L. J. and Bracker, C. E., 1972. *Protoplasma*, 74: 271—305.
Livne, A., 1964. *Pl. Physiol., Lancaster*, 39: 614—621.
Livne, A. and Daly, J. M., 1966. *Phytopathology*, 56: 170—175.
Long, D. E. and Cooke, R. C., 1974. *New Phytol.*, 73: 889—899.
Long, D. E., Fung, A. K., McGee, E. E. M., Cooke, R. C. and Lewis, D. H., 1975. *New Phytol.*, 74: 173—182.
Lunderstädt, J., Heitefuss, R. and Fuchs, W. H., 1962. *Naturwissenschaften*, 49: 403.
Macdonald, P. W. and Strobel, G. A., 1970. *Pl. Physiol., Lancaster*, 46: 126—135.
Maclean, D. J., 1974. *Trans. Br. mycol. Soc.*, 62: 333—349.
Maclean, D. J. and Scott, K. J., 1970. *J. gen. Microbiol.*, 64: 19—27.
Maclean, D. J., Scott, K. J. and Tommerup, I. C., 1971. *J. gen. Microbiol.*, 65: 339—342.
McKeen, W. E., Smith, R. and Bhattacharya, P. K., 1969. *Can. J. Bot.*, 47: 701—706.
Maheshwari, R., Hildebrandt, A. C. and Allen, P. J., 1967. *Bot. Gaz.*, 128: 153—159.
Manocha, M. S., 1975. *Phytopath. Z.*, 82: 207—215.
Manocha, M. S. and Shaw, M., 1967. *Can. J. Bot.*, 45: 1575—1582.
Manocha, M. S. and Wisdom, C. J., 1971. *Phytopath. Z.*, 70: 263—273.
Millerd, A. and Scott, K. J., 1962. *A. Rev. Pl. Physiol.*, 13: 559—574.
Mirocha, C. J. and Rick, P. D., 1967. In C. J. Mirocha and I. Uritani (eds.), *The Dynamic Role of Molecular Constituents in Plant—Parasite Interactions*, p. 121—141, Bruce, St. Paul, Minnesota.
Mirocha, C. J. and Zaki, A. I., 1965. *Phytopathology*, 55: 940—941.
Mirocha, C. J. and Zaki, A. I., 1966. *Phytopathology*, 56: 1220—1224.
Mitchell, S. and Shaw, M., 1969. *Can. J. Bot.*, 47: 1887—1889.
Mount, M. S. and Ellingboe, A. H., 1969. *Phytopathology*, 59: 235.
Person, C. O., 1960. *Can. J. Genet. Cytol.*, 2: 103—104.
Peyton, G. A. and Bowen, C. C., 1963. *Am. J. Bot.*, 50: 787—797.
Pozsár, B. I. and Király, Z., 1966. *Phytopath. Z.*, 56: 297—309.
Quick, W. A. and Shaw, M., 1964. *Can. J. Bot.*, 42: 1531—1540.
Rajendren, R. B., 1972a. *Bull. Torrey bot. Club*, 99: 84—88.
Rajendren, R. B., 1972b. *Mycologia*, 64: 591—598.
Reddy, M. N. and Williams, P. H., 1970. *Phytopathology*, 60: 1463—1465.
Robb, J., Harvey, A. E. and Shaw, M., 1975a. *Physiol. Pl. Path.*, 5: 1—8.
Robb, J., Harvey, A. E. and Shaw, M., 1975b. *Physiol. Pl. Path.*, 5: 9—18.
Rohringer, R. and Heitefuss, R., 1961. *Can. J. Bot.*, 39: 263—267.
Rohringer, R., Samborski, C. J. and Person, C. O., 1961. *Can. J. Bot.*, 39: 775—784.
Sargent, J. A., Tommerup, I. C. and Ingram, D. S., 1973. *Physiol. Pl. Path.*, 3: 231—240.
Savile, D. B. O., 1971. *Q. Rev. Biol.*, 46: 211—218.
Schipper, A. L. and Mirocha, C. J., 1969a. *Phytopathology*, 59: 1416—1422.
Schipper, A. L. and Mirocha, C. J., 1969b. *Phytopathology*, 59: 1722—1727.
Schnepf, E., Hegewald, E. and Soeder, C. J., 1971. *Arch. Mikrobiol.*, 75: 209—229.

Schnepf, E., 1972. *Protoplasma*, 75: 155—165.

Scott, K. J., 1965. *Phytopathology*, 55: 438—441.

Scott, K. J., 1972. *Biol. Rev.*, 47: 537—572.

Scott, K. J. an'd Maclean, D. J., 1969. *A. Rev. Phytopath.*, 7: 123—146.

Scott, K. J. and Smillie, R. M., 1963. *Nature, Lond.*, 197: 1319—1320.

Scott, K. J. and Smillie, R. M., 1966. *Pl. Physiol., Lancaster*, 41: 289—297.

Sequeira, L., 1963. *A. Rev. Phytopath.*, 1: 1—30.

Sequeira, L., 1973. *A. Rev. Pl. Physiol.*, 24: 353—380.

Shaw, M., 1963. *A. Rev. Phytopath.*, 1: 259—294.

Shaw, M. and Hawkins, A. R., 1958. *Can. J. Bot.*, 36: 1—16.

Shaw, M. and Manocha, M. S., 1965. *Can. J. Bot.*, 43: 1285—1292.

Shaw, M. and Samborski, D. J., 1957. *Can. J. Bot.*, 35: 389—406.

Shaw, M. D., Samborski, D. J. and Oaks, A., 1958. *Can. J. Bot.*, 36: 233—237.

Siddiqui, M. A. and Manners, J. G., 1971. *J. exp. Bot.*, 22: 792—799.

Slesinski, R. S. and Ellingboe, A. H., 1971. *Can. J. Bot.*, 49: 303—310.

Smith, D., Muscatine, L. and Lewis, D., 1969. *Biol. Rev.*, 44: 17—90.

Sommer, N. F., 1961. *Physiologia. Pl.*, 14: 460—469.

Srivasta, B. I. S. and Shaw, M., 1962. *Can. J. Bot.*, 40: 309—315.

Staples, R. C. and Ledbetter, M. C., 1960. *Contr. Boyce Thompson Inst. Pl. Res.*, 20: 349—351.

Sziráki, I., Balázs, E. and Király, Z., 1975. *Physiol. Pl. Path.*, 5: 45—50.

Temmink, J. H. M. and Campbell, R. N., 1969. *Can. J. Bot.*, 47: 421—424.

Thornton, J. D. and Cooke, R. C., 1970. *Trans. Br. mycol. Soc.*, 54: 361—365.

Thornton, J. D. and Cooke, R. C., 1974. *Physiol. Pl. Path.*, 4: 117—126.

Thrower, L. B., 1965a. *Phytopath. Z.*, 52: 269—294.

Thrower, L. B., 1965b. *Phytopath. Z.*, 52: 319—334.

Thrower, L. B. and Lewis, D. H., 1973. *New Phytol.*, 72: 501—508.

Thrower, L. B. and Thrower, S. L., 1966. *Phytopath. Z.*, 57: 267—276.

Tiwari, M. M. and Arya, H. C., 1969. *Science N.Y.*, 163: 291—293.

Trocha, P. and Daly, J. M., 1970. *Pl. Physiol., Lancaster*, 46: 520—526.

Turel, F. L. M., 1969. *Can. J. Bot.*, 47: 821—823.

Turel, F. L. M., 1971. *Can. J. Bot.*, 49: 1993—1997.

Van Dyke, C. G. and Hooker, A. L., 1969. *Phytopathology*, 59: 1934—1946.

Van Sumere, C. F., Van Sumere-De Preter, C. and Ledingham, G. A., 1957. *Can. J. Microbiol.*, 3: 761—770.

Wang, D., 1961. *Can. J. Bot.*, 39: 1595—1604.

Whitney, H. S., Shaw, M. and Naylor, J. M., 1962. *Can. J. Bot.*, 40: 1533—1544.

Williams, P. G., 1975. *Trans. Br. mycol. Soc.*, 64: 15—22.

Williams, P. G. and Hartley, M. J., 1971. *Nature, (New Biology) Lond.*, 229: 181—182.

Williams, P. G., Scott, K. J. and Kuhl, J. L., 1966. *Phytopathology*, 56: 1418—1419.

Williams, P. G., Scott, K. J., Kuhl, J. L. and Maclean, D. J., 1967. *Phytopathology*, 57: 326—327.

Williams, P. H., 1966. *Phytopathology*, 56: 521—524.

Williams, P. H., Aist, S. J. and Aist, J. R., 1971. *Can. J. Bot.*, 49: 41—47.

Williams, P. H., Aist, J. R. and Bhattacharya, P. K., 1973. In R. J. W. Byrde and C. V. Cutting (eds.), *Fungal Pathogenicity and the Plant's Response*, p. 141—155, Academic Press, London.

Williams, P. H., Keen, N. T., Strandberg, J. O. and McNabola, S. S., 1968. *Phytopathology*, 58: 921—928.

Williams, P. H. and McNabola, S. S., 1967. *Can. J. Bot.*, 45: 1665—1669.

Williams, P. H. and McNabola, S. S., 1970. *Phytopathology*, 60: 1557—1561.

Williams, P. H. and Pound, G. S., 1964. *Phytopathology*, 54: 446—451.

Williams, P. H. and Yukawa, Y. B., 1967. *Phytopathology*, 57: 682—687.

Wolf, F. T., 1974. *Can. J. Bot.*, 52: 767—772.

Wong, A. L. and Willetts, H. J., 1970. *Trans. Br. mycol. Soc.*, 55: 231—238.

Wood, R. K. S., 1967. *Physiological Plant Pathology*, Blackwell, Oxford and Edinburgh.

Wynn, W. K., 1963. *Phytopathology*, 53: 1376—1377.

Yarwood, C. E., 1967. *A. Rev. Pl. Physiol.*, 18: 419—438.

Yarwood, C. E. and Jacobson, L., 1955. *Phytopathology*, 45: 43—48.

Zimmer, D. E., 1970. *Phytopathology*, 60: 1157—1163.

Chapter 10

Obligate Necrotrophs and Hemibiotrophs

Facultative necrotrophs and obligate biotrophs exhibit two distinct and extreme kinds of behaviour. The necrotrophs destroy their hosts relatively rapidly and are able to exist, normally very successfully, as free-living saprotrophs. The biotrophs achieve a degree of balance with their hosts, modify and control host physiology, often in very subtle ways, and usually have no capacity for a free-living existence in nature. Facultative necrotrophs and obligate biotrophs thus comprise two relatively compact and sharply delimited groups each being defined by a number of distinct common characteristics. However, between the two extremes represented by them lie a great number of fungi which do not sit easily in either group. In many instances this may be due to lack of knowledge concerning their general biology, but there are some well-investigated fungi that are clearly either necrotrophs or biotrophs yet have very different characteristics from typical representatives of these two categories. These species may, albeit tentatively, be divided into two rather ill-defined series; obligate necrotrophs and hemibiotrophs.

Examples of fungi in these series are listed in Table 13 but two points concerning this list require emphasis. First, it represents only a selection of such species and it is possible to include many others, although these might be rather less familiar than those that are included. Second, there is so little information concerning certain aspects of their biology that it is entirely possible that in the future species will be relocated in other symbiotic groupings. Before discussing their salient features it should be noted that, while some of these fungi have properties that place them in intermediate positions between facultative necrotrophy and obligate biotrophy, they do not necessarily form a link, in evolutionary terms, between these two extremes. It has been argued that biotrophy developed from necrotrophy but it is equally possible that each mode of life arose and evolved independently (Lewis 1973, 1974). This can also be said of obligate necrotrophy and hemibiotrophy, although it is tempting to use the examples from these two series to bridge the distance between facultative necrotrophs and

Table 13
Examples of obligate necrotrophs and hemibiotrophs

Species causing massive tissue destruction	Species causing localized damage
Obligate necrotrophs	**Obligate necrotrophs**
Some species within '*Armillaria mellea*'	*Botrytis fabae*
Heterobasidion annosum	*Gaeumannomyces graminis*
Some *Botrytis* species, *B. allii*	Numerous leaf-spotting and stem-
Many *Sclerotinia* and *Sclerotium*	spotting fungi, mainly
species, *Sclerotinia fructicola*,	Ascomycotina and
Sclerotinia fructigena, *Sclerotium*	Hyphomycetes
cepivorum	
Some damping-off fungi especially	**Hemibiotrophs**
Fusarium species	
Some soil-borne species of *Phytophthora*	*Phytophthora infestans* on some
and *Pythium*	solanaceous hosts
	Rhynchosporium secalis
Fungi which initially occupy the	*Venturia inaequalis*,
vascular tract and are later necro-	*V. rumicis*, *V. pirina*
trophic on moribund host tissues.	*Guignardia bidwellii*
Fusarium and *Verticillium* species	*Mycosphaerella* species
	Numerous leaf-spotting and
Hemibiotrophs	stem-spotting fungi, mainly
Phytophthora infestans on some	Ascomycotina and
solanaceous hosts	Hyphomycetes

obligate biotrophs. All four modes may be ends in themselves, one not necessarily giving rise to or preceding another.

1. Obligate Necrotrophs

Some obligate necrotrophs, like facultative necrotrophs, rapidly colonize their hosts and death of the latter is accompanied by massive tissue destruction. Other obligate necrotrophs, however, cause only localized damage to their hosts which then become unthrifty but die, if at all, only slowly. In either case death of the host or that part of the host which is occupied terminates necrotrophic activity and a purely saprotrophic phase then ensues. In contrast to facultative necrotrophs, obligate necrotrophs have a poorly-developed capacity for a free-living saprotrophic existence under natural conditions. Since, without exception, they can be grown on relatively simple media in axenic culture this low capacity is due to an inability to compete with other micro-organisms during saprotrophic colonizaiton of dead organic materials rather than being determined by specialized nutrient require- ments. Their saprotrophic activities for the most part are, therefore, virtually restricted to the dead tissues which they have themselves created during necrotrophic colonization of the host (Garrett, 1970). Within these dead organic substrates obligate necrotrophs are able to survive until new hosts become available for infection. Because their saprotrophic phase is so severely limited, and since the provision of substrates within which they can survive depends on a preceding necrotrophic phase, they are thus obligately symbiotic, requiring association with a host for their normal phenotypic development. Obligate necrotrophs survive in a variety of ways within their dead substrates, frequently existing as either limited vegetative mycelia, resting mycelia, spores, or sclerotia all of which remain *in situ* and infect hosts that enter the zone that they occupy. Some species produce mycelial strands or rhizomorphs which allow them to grow away from dead substrates, until a suitable host is contacted, by using the former as food bases to provide energy for growth.

Although being in possession of dead organic matter confers an advantage on these fungi, during their saprotrophic phase invasion of their substrates by other fungi is inevitable and the period during which obligate necrotrophs can survive, at least in an actively growing mycelial form, will depend on the extent and rapidity with which colonization by more competitive saprotrophs takes place. If the saprotrophic phase of obligate necrotrophs is taken to be strictly that period during which active hyphal growth is taking place in the tissues of their dead host, then it is obvious that, while the saprotrophic phase of facultative necrotrophs is potentially of indefinite duration, that of obligate necrotrophs is finite. Depending on the organism involved, and prevailing environmental conditions, this period may be relatively long, but even so, there is usually no movement from occupied tissues into other neighbouring dead organic substrates. It is, however, sometimes more difficult than it might seem to draw a distinction between obligate and facultative necrotrophs with respect to their saprotrophic phases. Again, depending on environmental conditions some species, while quite justifiably being termed obligate necrotrophs, do have some ability to colonize dead non-host substrates, and similarly, in some situations, not all facultative necrotrophs are successful and efficient competitive saprotrophs. The line of demarcation between the two groups is thus commonly blurred. For example, the antagonistic activities of foot-rotting and root-rotting *Fusarium* species, many damping-off and root-invading fungi, including species of *Phytophthora* and *Pythium*, vascular-wilt fungi, and the ubiquitous *Rhizoctonia solani* are all well known and are documented in great detail. Many of these necrotrophs can spread rapidly through the soil from host to host when these are in close proximity to one another, use dead host remains as food bases for further infections and many can undoubtedly survive, often for considerable periods, in dead organic substrates (Gibson, 1956; Garrett, 1962; Long and Cooke, 1969; Garrett, 1970). Yet their ability for active saprotrophic colonization of already dead organic material in the face of competition from other soil fungi has been infrequently studied, and circumstantial evidence points to this ability being low in a great many species. It is highly probable that a number of these highly successful fungi are ecologically obligate necrotrophs although a whole range of saprotrophic ability obviously exists among them (Rao, 1959; Stover, 1962; Vujičić and Park, 1964; Banihashemi and Dezeeuw, 1973; El-Abyad and Saleh, 1973; Garrett, 1975).

Host Recognition

Contact between the actively growing mycelium or sedentary propagules of a soil-borne obligate necrotroph and a susceptible host organ, usually a root, will obviously depend to a great extent on chance, but some obligate necrotrophs have evolved mechanisms by means of which this random element is reduced or even totally eliminated. These fungi either have a general capacity to respond positively to the proximity of living roots of both host and non-host species, or show a specialized response to the roots of host species only. In the first case wastage of propagules or reduction in inoculum potential of mycelium, through random germination or random growth, is diminished, although a great deal of wastage will still occur through responses to non-host roots. However, within the root system of a single host or the intermingled roots of a host monoculture this mechanism aids the rapid spread and establishment of the fungus. In the second case the highly specific response to suitable hosts reduces wastage to a minimum through accurate 'recognition' and location of the host by the fungus.

In common with all other soil micro-organisms obligate necrotrophs are affected by root exudates diffusing into the rhizophere. Such diffusates may accelerate mycelial growth, stimulate spore or sclerotium germination, and cause orientated growth of germ tubes, or germling hyphae arising from germinating propagules, to take place towards the root along a diffusion gradient (Rovira, 1965; Garrett, 1970; Coley-Smith and Cooke, 1971). In addition, when zoospores are produced these often show positive chemotaxis with respect to roots, moving towards them and accumulating upon them (Royle and Hickman, 1964; Hickman and Ho, 1966). Propagule germination, chemotropism and chemotaxis can be evoked by a number of host metabolites which diffuse into the rhizosphere, for instance, sugars, amino acids, organic acids and ethanol (Chang-Ho and Hickman 1970; Hickman, 1970; Allen and Newhook, 1973). Some of these compounds produce their effect by acting as major exogenous nutrients for hyphae and propagules.

Chemotactic responses of zoospores are brought about by changes in the orientation of their flagella (Allen and Newhook, 1973; Khew and Zentmeyer, 1974). Organic compounds commonly diffusing from plant roots, particularly ethanol, alter this orientation so that swimming is directed along a concentration gradient of the compound or compounds concerned. This response will occur whether or not the source of the compound is a host or a non-host root. This means that the chemically-induced responses of many obligate necrotrophs are unrelated to the breadth or narrowness of their host range, and are the same irrespective of whether the rhizosphere which they inhabit is that of a host or a non-host species, although successful infection will obviously only occur after contact with host roots (Goode, 1956; Coley-Smith and Cooke, 1971).

There are notable exceptions to this generalization and more specific reactions to rhizosphere diffusates do occur (Rovira, 1965). These are usually of two kinds. First, there may be a lack of response to the presence of non-host roots which is due to the production of growth or germination inhibitors by those roots. Second, germination, chemically-directed growth, or both, may take place only in response to compounds or combinations of compounds peculiar to the roots of a particular group, sometimes a very restricted group, of host plants (Kerr and Flentje, 1957). Both types of response are typical of some fungi with relatively narrow host ranges and, in the second kind, highly specific host—fungus interactions occur during the pre-symbiotic phase before contact between the two organisms has been made. Perhaps the best known example of such an interaction is that between the sclerotium-forming soil-borne fungus *Sclerotium cepivorum* and its hosts (King and Coley-Smith, 1968; Coley-Smith and King, 1969; King and Coley-Smith, 1969). *S. cepivorum* is restricted to the genus *Allium* and survives in the soil in the form of sclerotia that, unlike many other kinds of sclerotia, are capable of germinating only once, and which do not normally germinate in non-sterile soil in the absence of a suitable host.

Germination is stimulated by *Allium* root diffusates, the hyphal plugs produced during germination showing orientated growth towards the host root. During the period between germination and contact with the host, the vegetative hyphae are extremely susceptible to antagonism and competition from other soil micro-organisms. The specific response to *Allium* root diffusates thus has an obvious survival value in ensuring that the single germination takes place only when suitable hosts are available for infection. The compounds responsible for inducing germination in the soil are volatile alkyl sulphides, particularly *n*-propyl and allyl sulphides, which are characteristic of *Allium* species, although they also sometimes occur in small quantities in crucifer species. These compounds are not evolved by intact *Allium* plants and are in any case rather insoluble in water. Intact plants do, however, exude water-soluble alkyl cysteine sulphoxides and in the soil these are metabolized by bacteria to yield stimulatory mixtures of volatile alkyl sulphides which then act on the sclerotia.

The Necrotrophic Phase

After contact with a suitable host has been made the necrotrophic phase of the host—fungus association is initiated. There is no evidence to suggest that the routes of entry or mechanisms by means of which obligate necrotrophs gain access to host tissues differ in any way from those that characterize facultative necrotrophs. Direct penetration of intact host surfaces, entry through wounds, root hairs, and juvenile root or hypocotyl tissues, all occur depending on the host—fungus combination. These processes, together with subsequent colonization of the host, are aided or brought about entirely by the production of an array of extracellular degradative enzymes identical in nature and mode of action with those characteristic of facultative necrotrophs. For instance, *Gaeumannomyces graminis* produces cellulolytic enzymes, pectinmethylesterase and polygalacturonases, and the ability of various isolates of this species to invade and exploit cereal roots is related to their ability to degrade cellulose (Weste, 1970a, b; Pearson, 1974).

Where putative obligate necrotrophs infect leaves, for example *Alternaria solani* on tomato, their effects on the carbon dioxide assimilation patterns of the host are similar to those brought about by facultative necrotrophs in that there is retention of photosynthates within the infected leaf but no increased import into that leaf from adjacent healthy ones (Coffey, Marshall and Whitbread, 1970). Infection of wheat by *Septoria nodorum* reduces assimilation rate per unit area of leaf, presumably through destruction of photosynthetic tissue, but there is no significant difference in the respiration rates found in diseased and healthy tissues. There is also evidence that a large proportion of photosynthate found in infected tissues is unavailable for export to other parts of the wheat plant (Sharen and Taylor, 1968).

Root infections can also bring about changes in the distribution of assimilates within the plant. Infection of the roots of wheat and barley by *Gaeumannomyces graminis* causes an increase in the movement of assimilates to the roots and premature development of the crown root system (Asher, 1972a). Net assimilation of whole plants is not affected by *G. graminis* infection unless this is severe, but respiratory losses are reduced by the presence of the fungus in root tissues. ^{14}C-labelled photosynthates which are translocated to the roots do not, as might be supposed, accumulate within or immediately around infection sites, but accumulate in the undamaged parts of infected root systems and in particular in the young crown roots (Asher, 1972b). Enhanced movement of assimilates to the root system does not, therefore, appear to be due to an increased demand for substrates by infected roots and seems to be not specifically associated with disease, since amputation of roots in healthy plants can also induce accumulation of labelled assimilates in crown roots. In addition, when plants are continuously supplied with water the proportion of labelled assimilate translocated to the root systems of healthy and infected plants in identical. It is likely that root infection by *G. graminis*, like root amputation, interferes with the uptake of water by the roots, so creating a water deficit in the shoot which, in turn, leads to inhibition of leaf growth. Assimilates are then diverted to those organs where active growth is still taking place, principally the crown roots.

2. Hemibiotrophs

During symbiosis the mode of nutrition of hemibiotrophic fungi passes through two distinct phases, being at first biotrophic then necrotrophic. During the biotrophic phase there may be little or no host cell wall degradation and host protoplasts remain alive (Mercer, Wood and Greenwood, 1975). In addition if, for example, an infected leaf dies and then falls from the host the necrotrophic phase is often followed by one of saprotrophy. Saprotrophic activity is, however, restricted to tissues that were occupied by the fungus when it was necrotrophic, so that hemibiotrophs have only a restricted or transient capacity for a free-living existence. This pattern of behaviour is exemplified by many leaf-spotting fungi of which the Pyrenomycete *Guignardia bidwellii* on *Vitis rotundifolia* is a typical representative (Luttrell, 1974). *G. bidwellii* infects young leaves and is biotrophic within juvenile tissues for a period of days or weeks. As the leaves mature further biotrophic colonization becomes restricted and the fungus becomes necrotrophic. The occupied areas of tissue die as a result of necrotrophic activity of the fungus which then produces pycnidia on the dead tissue. In the Autumn, after leaf fall, the fungus begins to spread again through leaf tissue, but this time it is saprotrophic on dead host cells and ascocarps are formed in the occupied tissues. During the saprotrophic phase the hemibiotroph is competing with other micro-organisms for available space and nutrients but has some advantage in that it is already in possession of substantial areas of leaf tissue.

Hemibiotrophy is widespread among Ascomycotina and Hyphomycetes that produce limited or late necroses on leaves, stems and fruit. Typically, asexual reproduction occurs during the necrotrophic phase with sexual reproduction taking place much later during the saprotrophic phase. It has been suggested that the Ascomycotina life cycle in some way has an inherent adaptability to the hemibiotrophic mode of life, and that hemibiotrophy is not a transient stage in the evolution of biotrophy but rather represents the end point of a particular developmental pathway which has been taken by some of these fungi (Luttrell, 1974). Hemibiotrophy is, however, by no means exclusive to the Ascomycotina and Hyphomycetes, and many other fungi, in particular some members of the Peronosporales, are quite clearly hemibiotrophs. One of these species, *Phytophthora infestans,* is usually treated, in discussions on the physiology of biotrophy, as a typical biotroph yet it has a distinct and important necrotrophic phase.

Hemibiotrophs differ from biotrophs in one further respect. They are normally easily grown in axenic culture on chemically defined media upon which, with the notable exception of *P. infestans,* their morphology generally remains more or less constant, and upon which their ability to subsequently infect suitable hosts is relatively undiminished even after subculture over long periods of time (Hall, 1959; Hodgson and Sharma, 1967; Caten, 1971). The ease with which they can be brought into and maintained in axenic culture, and the lack of variability which the majority show while in culture, contrasts with the difficulties encountered in attempts to grow many biotrophs.

Although it is probable that many hundreds of species are hemibiotrophs, detailed investigations on the physiology of host—fungus relationships during the biotrophic and necrotrophic phases have been infrequently made, and these have been mainly confined to three fungi; *Venturia inaequalis* on apple, *Rhynchosporium secalis* on barley and *Phytophthora infestans* on potato. *V. inaequalis* and *R. secalis* can be looked upon as being typical fruit- and leaf-spotting hemibiotrophs, but even so it is by no means certain that their behaviour during symbiosis reflects a general pattern to be found among all other fungi of this kind. The information available on these two species does, however, represent a beginning in the study of this group, and indicates to what degree they differ from biotrophs. *P. infestans* stands apart from *V. inaequalis* and *R. secalis* not only phylogenetically but also in terms of its effect on its host. At the same time it has behavioural characteristics that make it impossible to consider it, as has been done constantly in the past, as a biotroph.

Rhynchosporium secalis and Venturia inaequalis

After infection of host leaves, development of both *R. secalis* and *V. inaequalis* is at first by means of the production of a mycelial mat between the outer wall of the epidermal cells and the overlying cuticle (Preece, 1963; Ayesu-Offei and Clare, 1970). However, there are considerable differences between the two fungi with respect to fungus—host relationships during this and subsequent stages of development.

In *R. secalis* the time elapsing between initial infection of barley leaves and the appearance of lesions is short, varying between 3—7 days. Within this period extensive subcuticular hyphal growth takes place and the lack of visible host response indicates that during this stage the nutrition of the fungus is biotrophic. Soluble nutrients in the free space of the leaf are utilized and there is an increase in the permeability of the underlying host cells, thus increasing the concentration of these nutrients (Jones and Ayres; 1972). Any damage to host cells which does occur is restricted to the epidermis where the cell walls lying below the mycelial mat become swollen, lamellate, and collapse, so bringing the inner and outer walls of the epidermal cells together. The fungus eventually becomes necrotrophic and pronounced symptoms subsequently appear when the mesophyll cells become affected. Those cells beneath the mycelial mat eventually collapse and die, but this occurs before they are entered by hyphae, and it is only these areas of dead cells that are subsequently entered. There is no direct lateral or longitudinal spread of the fungus within deeper leaf tissue, lesions increasing in diameter as a result of the marginal growth of the overlying subcuticular mycelium. Sporulation then takes place within the necrotic areas (Ayesu-Offei and Clare, 1970; Fowler and Owen, 1971). Why *R. secalis* should be biotrophic when subcuticular but necrotrophic within the mesophyll is not clear, but this behaviour suggests that it has a low degree of compatibility with living host cells, so that a physiological balance between the two organisms, of the kind which is typical of biotrophy, is not possible once the fungus has passed a certain point in its subcuticular development. This may be due to the uncontrolled action of its degradative enzymes. *R. secalis* has been shown to produce a number of cellulolytic enzymes *in vitro*, and the collapse of cell walls during lesion growth indicates that these, and other cell wall degrading enzymes, are also produced *in vivo* (Olutiola and Ayres, 1973). Although the symptomless, subcuticular, biotrophic stage is quickly followed by one of pronounced necrotrophy, the mycelium at the advancing edge of the necrotic lesion is still exclusively subcuticular and is, presumably, behaving biotrophically. It is not yet known whether the biotrophic mycelium of *R. secalis* has the same effects on host physiology as those that are well documented for obligate biotrophs. However, the occurrence of extensive chlorosis around lesions suggests that the synthetic processes of the host are unlikely to be affected in a positive way, and that the total effect of infection by this particular hemibiotroph may be equivalent to that found in necrotrophic associations. In this regard there is some preliminary evidence that there is retention of metabolites within infected areas, but that there is neither an increase in import into affected tissues, nor an increase in synthesis in or around infection sites.

Only young apple leaves are susceptible to infection by *Venturia inaequalis*, and there may be a period of several weeks between infection and the appearance of visible symptoms. During this time extensive subcuticular mycelium is produced and asexual reproduction may take place on the leaf surface (Preece, 1963). As with *R. secalis*, the subcuticular growth stage is biotrophic but in the case of *V. inaequalis* this phase is of much longer duration. As infection progresses, and sporulation increases, the cuticle above the mycelium become raised and ruptured, with pigment deposition and death of the host's epidermal cells taking place. The fungus is then presumably behaving as a necrotroph within these areas, although subcuticular hyphae growing at the margin of the lesion will still be biotrophic. In contrast to infections by *R. secalis*, the mesophyll cells are not invaded until late in the growing season of the host so that *V. inaequalis* generally only moves from its subcuticular position into deeper leaf tissues shortly before leaf fall. Perithecial primordia are then formed within the mesophyll and complete their development after over-wintering on fallen leaves. Nutrition of the fungus in the mesophyll is thus finally, saprotrophic.

Again the question is raised, as with *R. secalis*, as to why this fungus is biotrophic only when occupying that part of the leaf between the cuticle and the epidermis, although in *V. inaequalis* the situation is rather different in that the subcuticular stage is of relatively long duration. It is possible that the host's epidermal cells physically restrict deeper invasion and that this, coupled with catabolite repression of the degradative enzymes of the fungus, determines both its position and nutritional behaviour. There is also evidence that physical barriers to deeper invasion are produced in the mesophyll as a result of infection. Some kind of physical balance with the host would prevail until damage to the cuticle during production of asexual spores caused severe injury to the underlying epidermis and, later, to the underlying mesophyll cells.

Despite being in a position outside the leaf tissues there is some evidence that *V. inaequalis* may markedly affect the movement of metabolites within the leaf (Hignett and Kirkham, 1967). Melano-proteins of molecular weight 10,000—70,000 from axenic culture filtrates, when injected together with ^{14}C-glucose into the petioles of healthy apple leaves, prevent the movement of label into the interveinal tissues, so that radioactivity is restricted to the vascular systems. If conidia labelled with ^{14}C-alanine are used to infect healthy leaves, then labelled metabolites derived from this material can subsequently be detected in the vascular system, and its distribution is similar to that obtained using petiole injections of melanoproteins plus labelled glucose. Although the evidence is not entirely conclusive, it has been suggested that in the host these melanoproteins are produced at infection sites from which they diffuse outwards into the leaf tissues. Here, their effect is to practically confine host translocates to the vascular network so that in some way a system of 'preferred routes' is established leading into the developing lesion. Export from the infected leaf is then progressively restricted and metabolites are thus made more available to lesion tissue. These routes seem to be established before the lesion becomes visible. Where the cuticle becomes damaged and lifted there is an increase in local transpiration rate which further increases movement of metabolites into the lesion, and at this point the 'preferred routes' disappear.

Other effects of pigmented products from culture filtrates of *V. inaequalis* have been noted (Kirkham and Hignett, 1970). Pigmented filtrates applied to healthy leaves together with conidia result in a 40% increase in the number of lesions produced when compared with those occurring on leaves inoculated with conidia alone. A similar but much smaller effect is obtained if kinetin is used instead of filtrate, but application of kinetin plus filtrate together with conidia results in a frequency of lesions almost identical to that produced by conidia alone. This, possibly, means that the effects of kinetin and filtrate are mutually antagonistic or mutually exclusive, their modes of action on the fungus, host, or both, being different from one another. The exact nature of the active fractions of pigmented extracts from *V. inaequalis* and details of their mode of action remain unknown.

Phytophthora infestans

Although a vast literature exists that is concerned with various aspects of the biology of *P. infestans* on solanaceous plants, little is known concerning details of host—fungus physiology. After infection of potato leaves, the nutrition of the fungus is at first biotrophic but hyphae quickly ramify through all the leaf tissues and within a short time, usually 2—3 days after infection, visible lesions appear that become necrotic at their centres, and which on susceptible hosts rapidly increase in size. Within the necrotic host tissues the fungus is necrotrophic, but at the margins of the developing lesions the fungus is growing biotrophically 3—6 mm in advance of the necrotic region (Farrell, 1971). As with *R. secalis* and *V. inaequalis*, biotrophy in *P. infestans* may be due to catabolite repression of its degradative enzymes. If this repression were only temporary, then necrosis would occur if repression were in some way alleviated.

If infected leaves are allowed to photosynthesize ^{14}C-labelled carbon dioxide then radioactive metabolites accumulate in high concentrations at the edges of necrotic lesions but not within them. Accumulation within tissues where lesions are developing but are not yet visible also occurs (Shaw and Samborski, 1956; Garraway and Pelletier, 1966; Farrell, 1971). Some photosynthate from healthy leaflets may move into infected leaflets but this appears to be due to passive import of

180

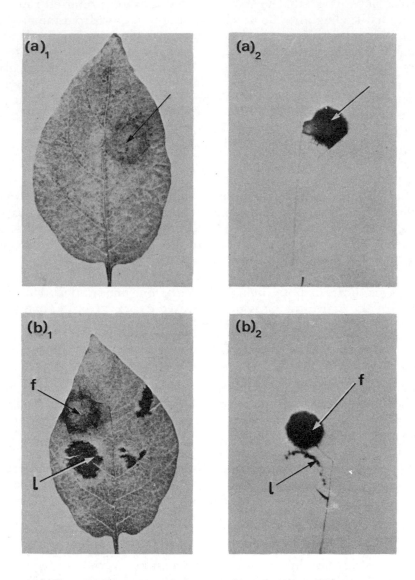

Figure 57 *Phytophthora infestans*. Distribution of ^{14}C-labelled photosynthates in uninfected and infected potato leaflets: (a), uninfected leaflet, position of ^{14}C-fed region arrowed; (a)$_2$ auto-radiograph of same leaflet showing direct movement of label to mid-vein; (b), infected leaflet, position of $^{14}CO_2$-fed region (f) and position of lesion (l) arrowed; (b)$_2$ autoradiograph of same leaflet showing accumulation of labelled photosynthate at margin of lesion. From Farrell, 1971; by permission of *Physiological Plant Pathology*

assimilates via the transpiration stream, and there is no good evidence for the occurrence of significant changes in the translocatery patterns of the host plant consequent upon infection. However, within an infected leaflet, if the lesion is in a suitable position, then photosynthate moving within uninfected tissue towards the mid vein accumulates within the infected but non-necrotic region (Figure 57). It is clear that accumulation and retention of assimilates occurs only in those host tissues in which *P. infestans* is biotrophic and, although no green island formation occurs at these sites, there is evidence for enhanced fixation of carbon dioxide within areas of biotrophic activity (Farrell, Preece and Wren, 1968; Farrell, 1971). Stomata within these areas are maintained permanently in an abnormally wide-open condition and show no significant diurnal variation in aperture. This alone could produce high assimilation rates in infected tissues with consequent accumulation of photosynthates. If stomatal opening in uninfected tissues is induced by their exposure to a carbon dioxide-free atmosphere, subsequent rates of assimilation by these tissues is comparable with those found in infected tissues. This suggests that the biochemical processes of photosynthesis are not altered by infection, but that enhanced assimilation is due solely to the removal of a physical restraint on gaseous exchange. What proportion of this assimilate is utilized by the fungus during its biotrophic phase is not known, and it is possible that the majority of it is used only when the fungus becomes necrotrophic within the assimilate-rich host tissues.

Production of sporangiophores only occurs at the margin of the necrotic zone, reflecting, perhaps, the dependence of *P. infestans* on living host tissues for the provision of factors essential for the sporulation process. Members of the Pythiaceae are unable to synthesize sterols but these compounds enhance mycelial growth of pythiaceous fungi and are also necessary for both their asexual and sexual reproduction (Hendrix, 1970). Sterols appear to be used as structural components for cell membranes and also act as precursors of hormonal and regulatory compounds within the hyphae (Langcake, 1975). *In vitro*, cholesterol, stigmasterol and β-sitosterol stimulate a 56%—80% increase in growth of *P. infestans* and promote sporangium formation. The same sterols are found in potato leaves (Langcake, 1974).

Bibliography

Allen, R. N. and Newhook, F. J., 1973. *Trans. Br. mycol. Soc.*, 61: 287—302.
Asher, M. J. C., 1972a. *Ann. appl. Biol.*, 70: 215—223.
Asher, M. J. C., 1972b. *Ann. appl. Biol.*, 72: 161—169.
Ayesu-Offei, E. N. and Clare, B. G., 1970. *Aust. J. Biol. Sci.*, 23: 299—307.
Banihashemi, Z. and Dezeeuw, D. J., 1973. *Trans. Br. mycol. Soc.*, 60: 205—210.
Caten, C. E., 1971. *Trans. Br. mycol. Soc.*, 56: 1—7.

182

Chang-Ho, Y. and Hickman, C. J., 1970. In T. A. Tousson, R. V. Bega and P. E. Nelson (eds.), *Root disease and Soil-Borne Pathogens*, p. 103—108, Berkeley, University of California Press.
Coffey, M. D., Marshall, C. and Whitbread, R., 1970. *Ann. Bot.*, 34: 605—615.
Coley-Smith, J. R. and Cooke, R. C., 1971. *A. Rev. Phytopath.*, 9: 65—92.
Coley-Smith, J. R. and King, J. E., 1969. *Ann. appl. Biol.*, 64: 289—301.
El-Abyad, M. S. and Saleh, Y. E., 1973. *Trans. Br. mycol. Soc.*, 60: 187—195.
Farrell, G. M., 1971. *Physiol. Pl. Path.*, 1: 457—467.
Farrell, G. M., Preece, T. F. and Wren, M. J., 1968. *Ann. appl. Biol.*, 63: 265—275.
Fowler, A. M. and Owen, H., 1971. *Trans. Br. mycol. Soc.*, 56: 137—152.
Garraway, M. O. and Pelletier, R. L., 1966. *Phytopathology*, 5: 1184—1189.
Garrett, S. D., 1962. *Trans. Br. mycol. Soc.*, 45: 115—120.
Garrett, S. D., 1970. *Pathogenic Root-Infecting Fungi*, Cambridge University Press.
Garrett, S. D., 1975. *Soil. Biol. Biochem.*, 7: 323—327.
Gibson, I. A. S., 1956. *E. Afr. agric. For. J.*, 21: 183—188.
Goode, P. M., 1956. *Trans. Br. mycol. Soc.*, 39: 357—377.
Hall, A. M., 1959. *Trans. Br. mycol. Soc.*, 42: 15—26.
Hendrix, J. W., 1970. *A. Rev. Phytopath.*, 8: 111—130.
Hickman, C. J., 1970. *Phytopathology*, 60: 1128—1135.
Hickman, C. J. and Ho, H. H., 1966. *A. Rev. Phytopath.*, 4: 195—220.
Hignett, R. C. and Kirkham, D. S., 1967. *J. gen. Microbiol.*, 48: 269—275.
Hodgson, W. A. and Sharma, K. P., 1967. *Can. J. Pl. Sci.*, 47: 447—449.
Jones, P. and Ayres, P. G., 1972. *Physiol. Pl. Path.*, 2: 383—392.
Kerr, A. and Flentje, N. T., 1957. *Nature, Lond.*, 179: 204—205.
Khew, K. L. and Zentmeyer, G. A., 1974. *Phytopathology*, 64: 500—506.
King, J. E. and Coley-Smith, J. R., 1968. *Ann. appl. Biol.*, 61: 407—414.
King, J. E. and Coley-Smith, J. R., 1969. *Ann. appl. Biol.*, 64: 303—314.
Kirkham, D. S. and Hignett, R. C., 1970. *Nature, Lond.*, 225: 388.
Langcake, P., 1974. *Trans. Br. mycol. Soc.*, 63 : 573—586.
Langcake, P., 1975. *Trans. Br. mycol. Soc.*, 64: 55—65.
Lewis, D. H. 1973. *Biol. Rev.*, 48: 261—278.
Lewis, D. H., 1974. In M. J. Carlile and J. J. Skehel (eds.), *Evolution in the Microbial World*, p. 367—392. 24th *Symp. Soc. gen. Microbiol.*, Cambridge University Press.
Long, P. G. and Cooke, R. C., 1969. *Trans. Br. mycol. Soc.*, 52: 49—55.
Luttrell, E. S., 1974. *Mycologia*, 66: 1—15.
Mercer, P. C., Wood, R. K. S. and Greenwood, A. D., 1975. *Physiol. Pl. Path.*, 5: 203—214.
Olutiola, P. O. and Ayres, P. G., 1973. *Trans. Br. mycol. Soc.*, 60: 273—282.
Pearson, V., 1974. *Trans. Br. mycol. Soc.*, 63: 199—202.
Preece, T. F., 1963. *Trans. Br. mycol. Soc.*, 46: 523—529.
Rao, A. S., 1959. *Trans. Br. mycol. Soc.*, 42: 97—111.
Rovira, A. D., 1965. In K. F. Baker and W. C. Snyder (eds.), *Ecology of Soil-Borne Plant Pathogens*, p. 170—186, Berkeley, University of California Press.
Royle, D. J. and Hickman, C. J., 1964. *Can. J. Microbiol.*, 10: 151—162.
Scharen, A. L. and Taylor, J. M., 1968. *Phytopathology*, 58: 447—451.
Shaw, M. and Samborski, D. J., 1956. *Can. J. Bot.*, 34: 389—405.
Stover, R. H., 1962. *Can. J. Bot.*, 40: 1473—1481.
Vujičić, R. and Park, D., 1964. *Trans. Br. mycol. Soc.*, 47: 455—458.
Weste, G., 1970a. *Phytopath. Z.*, 67: 189—204.
Weste, G., 1970b. *Phytopath. Z.*, 67: 327—336.

Mutualistic Symbioses with Plants

Chapter 11

Ectomycorrhizas

Numerous kinds of mutualistic associations exist between biotrophic fungi and the absorbing organs of a wide spectrum of vascular and non-vascular plants. The physical characteristics of these associations and the physiological consequences of symbiosis are very diverse and are in many cases far from fully understood (Meyer, 1966; Harley 1969, 1971; Meyer, 1974). On the basis of strictly morphological and anatomical features mycorrhizas can be divided into three broad groups (Peyronel, Fassi, Fontana and Trappe, 1969). First, there are ectomycorrhizas, in which the absorbing organ is entirely surrounded by a well-developed, usually compact, mantle of fungal material from which hyphae arise that pass into the organ and grow between its cells. Second, there are endomycorrhizas, where the hyphae external to the absorbing organ are not aggregated to any great extent, and in which intracellular as well as intercellular penetration by hyphae is characteristic. Finally, there are ectendomycorrhizas that share some of the features of both ecto- and endomycorrhizas. These have an external mantle of some sort, although it may not always be very well developed, and the hyphae within the host penetrate its cells as well as growing between them. Ectomycorrhiza, endomycorrhiza and ectendomycorrhiza correspond to the older and still commonly used terms *ectotrophic, endotrophic* and *ectendotrophic* mycorrhiza. Cogent and convincing arguments have been put forward that these older terms now be abandoned (Lewis, 1973). The basis of this proposal is that use of the suffix *—trophic* means that the names embody an erroneous implication that the three kinds of mycorrhiza obtain their nutrients in three distinct ways. That is literally by *outside, inside,* and *outside— inside* feeding respectively. Not only are these terms themselves nonsensical, and therefore have no precise applicability, but they also suggest the existence of uniform nutritional behaviour patterns within each group. This is patently not the case, various and diverse kinds of behaviour patterns being found within both endomycorrhizas and ectendomycorrhizas. The newer terminology, which accurately describes the physical organization of a particular mycorrhiza without suggesting that it has a particular function, is now rapidly coming into wider use (Hacskaylo, 1971). It should, however, be stressed that while ectendomycorrhizas are intermediate in structure between ecto- and endomycorrhizas, gradations occur between this intermediate type and both ectomycorrhizas on the one hand and endomycorrhizas on the other.

1. Features of Infection

It has been estimated that ectomycorrhizas occur in about 3% of the total number of plant species, with the great majority being found in forest trees (Meyer, 1966). A large number of both angiospermous and coniferous trees have ectomycorrhizas, but to date most experimental investigations on these symbioses have been concerned with either the European beech, *Fagus sylvaticus*, or various species of *Pinus*. Although there are variations in structure, all typical ectomycorrhizas have a number of common characteristics (Marks and Foster, 1973). The most obvious of these is the possession of a pseudoparenchymatous sheath of fungal material which entirely encloses the host root, including the apex, and which may comprise up to 40% of the total dry weight of the mycorrhizal organ. Hyphae or hyphal strands may be present which ramify from the sheath into the surrounding soil, though these are not a constant feature of all ectomycorrhizas. From the sheath numerous hyphae pass between the cells of the outer cortical tissue but rarely penetrate them. The outermost cortical cells may become transversely elongate, presumably as a result of contact with the fungus, and root hairs are absent. Both meristematic and root cap tissue are much reduced in comparison with those of non-mycorrhizal roots. It is usually lateral or short roots that become obviously mycorrhizal, and although main axes or mother roots may also be infected they do not always have an obvious or well-developed sheath and their apices remain fungus-free. Young laterals become infected as they emerge through the cortex of their infected mother roots and form profusely-branched systems, each branch being of limited length, that normally occupy the surface humus layers of the soil. Unlike their mother roots the short, mycorrhizal roots are relatively short-lived and are periodically replaced.

Wide diversity in the morphology and anatomy of ectomycorrhizas occurs, and while some of this variation can be accounted for in terms of different stages in development having markedly different characteristics, other factors are also involved. Possibly the most important determinant of ectomycorrhizal structure is the species of fungus that is associated with any particular host. Since more than one fungus may be involved in mycorrhiza formation on the root system of a single tree, mycorrhizal morphology, even on an individual host, may not be constant (Meyer, 1966). Several schemes with varying degrees of complexity have been devised in order to classify ectomycorrhizas on the basis of their structure, the aim being to define mycorrhizal groups in which small numbers of fungi, or even single species, are involved in mycorrhiza formation (Dominik, 1959; Chilvers and Pryor, 1965; Harley, 1969; Zak, 1973). It is, however, not known to what extent differences in structure may determine differences, if any, in physiology.

The majority of fungi so far known to form ectomycorrhizas are Basidiomycotina that belong to families within the Agaricales, although some are Gasteromycetes and, possibly, Ascomycotina (Trappe, 1962), the three major agaric genera being *Amanita, Boletus (Suillus)*, and *Tricholoma*. Ectomycorrhizal fungi when symbiotic are biotrophic but have closely related free-living saprotrophic counterparts which inhabit the same regions of the soil, that is the litter or humus layers, but which never form symbioses with plant roots. It is possible that ectomycorrhizal fungi evolved from such saprotrophs, and the former can be divided into four ecological groups depending on their ability for a free-living existence and on their degree of host specificity (Meyer, 1963). First, there are species, for example the Gasteromycete *Phallus impudicus*, that are normally free-living saprotrophs but which can also form mycorrhizas with suitable hosts. Second, there are fungi, for instance another Gasteromycete *Scleroderma aurantium*, that are normally mycorrhizal and have a broad host range, but which also have a well-developed ability for a free-living saprotrophic mode of life. The third group comprises the majority of known ectomycorrhizal species. These have little or no ability to live unassociated with plant roots but have a wide host range. Finally, there are a few fungi that are similarly not normally free-living and which, in addition, have extremely narrow host ranges. Some of these, for example *Boletus elegans* on *Larix*, may be restricted to a single host genus or species. It is from studies of associations involving members of the latter two groups, ecologically obligately mycorrhizal fungi, that the bulk of knowledge concerning the physiology of ectomycorrhizas has been obtained.

Many ecologically obligate ectomycorrhizal species can be brought into axenic culture with relative ease and the nutritional requirements, particularly carbon requirements, of a number of them have been determined. (Norkrans, 1949, 1950; Palmer and Hacskaylo, 1970; Hacskaylo, 1973). In general, the most favourable carbon sources for growth in axenic culture are glucose and other simple sugars. The more complex the carbohydrate the poorer mycelial growth is, and some species cannot utilize even disaccharides. It cannot, however, be said that all mycorrhizal fungi rely exclusively on a supply of simple carbohydrates, since a number of species have been found to utilize polysaccharides, particularly if they are also supplied with a small quantity of glucose, and some undoubtedly mycorrhizal fungi produce cellulases, hemicellulases and amylases in culture (Lamb, 1974). The ability to produce these exo-enzymes varies widely and, with very few exceptions, productivity is not as great as that found in free-living

Basidiomycotina that inhabit leaf litter (Lindeberg, 1948; Lyr, 1963). As a group, therefore, the obligately ectomycorrhizal fungi are probably unable to rapidly decompose litter or other forms of soil organic matter. Even if some of them have a limited ability to do so they are probably adversely affected by competition from more efficient litter-inhabiting fungi. Since simple sugars are the most ephemeral of carbon sources in soil the mycorrhizal fungi must depend largely, many of them probably entirely, on their hosts for such carbon compounds.

2. Carbohydrate Physiology

Using artificially infected seedlings of *Pinus sylvestris* it has been conclusively demonstrated that ^{14}C-labelled photosynthates are translocated from the shoots to mycorrhizal roots where they accumulate in the fungal mantle. There is also evidence that photosynthates and other metabolites may be transferred from one individual of *Pinus taeda* to another via interconnected ectomycorrhizas (Melin and Nilsson, 1957; Reid and Woods, 1969). A similar situation has been found in some endomycorrhizas of angiospermous trees, for example in *Acer* (Woods and Brock, 1964). Not only is there translocation into the sheath, but the rate of movement of photosynthate in *Pinus* species is greater to mycorrhizal roots than to non-mycorrhizal roots under equivalent conditions (Shiroya, Lister, Slankis, Krotkov and Nelson, 1962; Lister, Slankis, Krotkov and Nelson, 1968). While it is possible that this increase could be partially due to enhanced photosynthetic rates in mycorrhizal plants, it has been established that it is influenced mainly by the fungus acting as a metabolic sink for the host's carbon compounds.

Present understanding of the carbohydrate metabolism of ectomycorrhizal roots is based almost wholly on studies on *Fagus sylvatica*, but it is reasonable to suppose that other ectomycorrhizas function in a broadly similar, if not identical, way. Excised *Fagus* mycorrhizas have been found to contain trehalose, mannitol, glycogen, glucose, fructose and sucrose, while non-mycorrhizal roots lack trehalose, mannitol and glycogen. Trehalose and mannitol are absorbed readily by mycorrhizal roots but slowly or not at all by non-mycorrhizal roots. Trehalose, mannitol and glycogen are, therefore, carbon compounds peculiar to the mycorrhizal condition (Lewis and Harley, 1965a, b). If glucose or fructose are fed to mycorrhizal roots then these sugars are incorporated into disaccharides and insoluble carbohydrates, but there is very little change in the amount of internal reducing sugars. However, when mycorrhizal roots are fed with ^{14}C-labelled glucose, fructose, or sucrose most label accumulates in trehalose, mannitol and glycogen (Table 14) (Harley and Jennings, 1958; Lewis and Harley, 1965b).

Table 14

Distribution of radioactivity in soluble sugars and glycogen in mycorrhizal roots of *Fagus sylvatica* after feeding with ^{14}C-glucose, ^{14}C-fructose, and ^{14}C-sucrose (after Lewis and Harley, 1965*b*)

Sugar supplied	Percentage of radioactivity in total carbohydrate		Percentage of radioactivity within soluble carbohydrates due to:				
	Soluble	Insoluble (mainly glycogen)	Glucose	Fructose	Sucrose	Trehalose	Mannitol
Glucose	41.2	58.8	5.1	0	16.9	56.4	11.6
Fructose	87.2	12.8	2.4	10.5	13.6	3.2	70.3
Sucrose	46.0	54.0	1.9	1.9	5.8	37.3	53.1

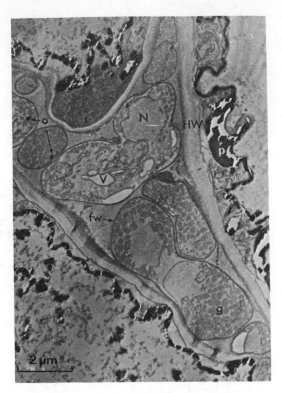

Figure 58 The ectomycorrhiza of *Pinus radiata*. Electronmicrograph of a transverse section through the outer cortex showing a group of hyphae lying between the cortical cells: fw, fungal cell wall; HW, host cell wall; N, fungal nucleus; V, vacuole; o, oil body in fungus; g, glycogen deposits within hypha; p, polyphenolic material. From Foster and Marks, 1966; by permission of *Australian Journal of Biological Sciences*

Large and important differences exist between the host tissue of mycorrhizal *Fagus* roots from which the sheath has been dissected, and the separated fungal tissue. Host tissue, and uninfected roots, synthesize sucrose from glucose or fructose while at the same time forming very little insoluble carbohydrate. In contrast, the fungal material synthesizes trehalose, mannitol and glycogen rather than sucrose. It should be noted that some glycogen synthesis can take place in tissues interior to the sheath, and in some *Pinus* species the intercellular hyphae lying within the cortical layers are glycogen-rich (Figure 58) (Marks and Foster, 1973). When intact mycorrhizal roots of *Fagus* are allowed to absorb exogenous sugars, sheath tissue takes up approxi-

Table 15
Distribution of radioactivity in the core and sheath of *Fagus*
mycorrhiza after translocation of ^{14}C-sucrose
(after Lewis and Harley, 1965c)[a]

Treatment	Percentage of total radioactivity	
	Core	Sheath
Sucrose supplied to host and fungus	24	76
Sucrose supplied to host only	32	68

[a] For diagram of method see Figure 59.

mately 70% of the carbohydrate supplied, with the core accounting for the remaining 30% (Table 15: Figure 59).

Trehalose, mannitol and glycogen are thus the major carbohydrates of the fungus while glucose, fructose and sucrose are the major carbohydrates provided by the host. Of the latter it is sucrose that is the main, if not only, translocatory carbohydrate in *Fagus*, ^{14}C-sucrose movement taking place predominantly within host tissues. Label then passes from these to the sheath where is accumulates in trehalose, mannitol and glycogen (Table 15; Figure 59). Since reciprocal flow of

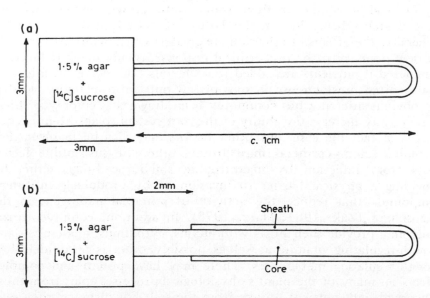

Figure 59   Experimental feeding of mycorrhizal *Fagus* roots with labelled sucrose: (a) intact root; (b) root with basal ring of fungal sheath removed. See also Table 15. From Lewis and Harley, 1965c; by permission of *New Phytologist*

carbon compounds from sheath to core does not occur, these fungal compounds, being non-utilizable by the host, act as a sink for host carbohydrates so that movement of these to the fungus is maintained. This mechanism clearly resembles that found in other mutualistic biotrophic fungi, for instance lichen fungi, and in antagonistic biotrophs. It should be emphasized that, while similar mechanisms probably operate in other ectomycorrhizas, that found in *Fagus* may differ from them in detail. In *Fagus* mycorrhiza, uptake of sucrose by the sheath involves its prior hydrolysis by the fungus, so that glucose and fructose are the sugars that are actually absorbed, the destination of glucose being trehalose and glycogen, that of fructose being mannitol (Lewis and Harley, 1965b). However, some mycorrhizal fungi utilize sucrose slowly or not at all *in vitro*, although adaptive enzymes for hydrolysis of this sugar may be produced after long exposure of the fungi to it (Palmer and Hacskaylo, 1970). This, together with the obvious consideration that sucrose may not always be the sole translocatory carbohydrate in trees other than *Fagus*, suggests that variations in ectomycorrhizal carbohydrate physiology are possible.

3. Effects on Host Growth

There is good evidence that, while not being absolutely dependent upon symbiosis, the growth of tree seedlings or cuttings is enhanced by mycorrhizal infection of their roots, both growth increments and mineral status being increased (Harley, 1969, 1970; Bowen, 1973). Generally, the effects of infection are greatest when plants are grown in nutrient-poor soils, the mycorrhizal response usually being reduced or eliminated if nutrients are added to such soils. The degree of infection is also often most intense in soils of low nutrient availability, so that the obvious inference has commonly been drawn that ectomycorrhizas in some way increase the ability of the root system to absorb nutrients. This is in fact the case, but other factors may also be involved. For example, ectomycorrhizas may protect otherwise susceptible feeder roots from infection by necrotrophic soil-borne fungi, either by providing a physical barrier to invasion or by producing antifungal compounds that reduce the activity of potential pathogens in the rhizosphere (Zak, 1964; Marx, 1973). In addition, ectomycorrhizal fungi may provide their hosts with auxins, cytokinins, gibberellins and growth-regulating vitamins as well as, possibly, other as yet unidentified growth-regulating metabolites. These may have potent and complex effects on many of the plant's physiological processes apart from those concerned with nutrient uptake from the soil. For instance, they may influence photosynthesis, translocatory patterns and morphogenesis, although their transfer from fungus to host remains to be proved (Slankis, 1973).

Analyses of mycorrhizal and non-mycorrhizal seedlings of *Pinus strobus* and *Pseudotsuga taxifolia* indicate that the mycorrhizal condition leads to a marked increase in phosphate absorption (McComb 1938, 1943; McComb and Griffith, 1946). Absorption of nitrogen and potassium is also enhanced, but this may be due to increased phosphorus absorption causing an increase in the general metabolic rate of host roots and hence the greater uptake of other elements. It is, perhaps, for this reason that the majority of studies on mineral uptake by ectomycorrhizas have been concerned with the absorption and fate of phosphorus (Bowen, 1973).

Since mycorrhizal short roots lack root hairs, the greater part of soil nutrients that pass into the host must first be absorbed by the outer layers of the sheath, then move through the sheath into the intercellular mycelium within the cortex. This mycelium probably provides a large contact surface between fungus and host which allows rapid nutrient exchange to take place. When present, the growth of hyphae or of mycelial strands into the soil from the sheath may result in extensive exploration of the soil around mycorrhizal roots by the mycorrhizal fungus, and it has been conclusively demonstrated that such extra-mycorrhizal mycelia can act in a manner analogous to root hairs with respect to mineral absorption. Labelled phosphorus, calcium and isotopic nitrogen have all been shown to move through hyphae and strands to the sheath in mycorrhizal pine seedlings and from there to the shoot system (Melin and Nilsson, 1950, 1953, 1955, 1958; Melin, Nilsson and Hacskaylo, 1958; Skinner and Bowen, 1974). However, extra-mycorrhizal mycelia are not essential for increased absorption by ectomycorrhizas and those devoid of them, for example in *Fagus* function equally efficiently (Harley, 1969). Obviously, the outer layers of sheath tissue must possess a well-developed ability to absorb minerals.

Uptake of phosphate by the mycorrhizal roots of *Pinus* species and of *Fagus* has been investigated in some detail. Phosphate uptake from solutions by both intact and excised mycorrhizal roots varies, but rates are approximately 2–5 times those found in comparable uninfected roots (Harley, 1970). In intact non-mycorrhizal roots of *Pinus radiata* there is an area of high phosphate absorption immediately behind the apex of the main axis corresponding to the zone of cell elongation. Additional sites of high uptake along the uninfected main axis are located at points where lateral short roots are emerging. Mycorrhizal main axes have a similar distribution of uptake sites, but those corresponding to the positions of emerging infected short roots are much more active than those in uninfected roots (Bowen, 1968). The phosphate absorbed by mycorrhizal short roots accumulates within the roots, and the primary site of accumulation is the fungal sheath. If *Fagus* mycorrhizas are excised and the sheath and core tissues separated by dissection then the sheath contains 5–8 times more phosphate on a tissue weight basis than does the core (Harley and McCready, 1952a, b). This demonstrates that in *Fagus*, as in *Pinus*, it is the sheath which is directly responsible for phosphate uptake and that it is unlikely that host cells, under the influence of the fungus, are imbued with greater absorbing power.

In *Fagus* mycorrhizas approximatley 90% of phosphate absorbed from solution remains in the sheath tissues, the remainder being found in the core. If, however, the sheath is first removed by dissection, then core tissues absorb at 4 times the rate found in core tissues of intact mycorrhizas (Harley and McCready, 1952b). The sheath is preventing the root tissues of the host from absorbing phosphate at the maximum rate possible, and so is exerting control on the host's phosphate uptake.

There are two possible routes for the movement of phosphate into sheath tissues from the soil, either through the interhyphal spaces or through an interior diffusion pathway provided by the fungal cells. At phosphate levels normally found in soils in nature, all phosphate probably moves through the diffusion pathway and little reaches the interior of the sheath by means of the interhyphal spaces. When *Fagus* mycorrhizas absorb ^{32}P-labelled phosphate, the first labelled compound detectable in the host is inorganic phosphate and the rate of arrival of this in host tissues is linear with respect to time. This demonstrates that during phosphate absorption by the mycorrhiza the phosphate passing to the host does not mix to any great extent, and has not previously been incorporated with, the inorganic phosphate of the sheath (Harley, Brierley and McCready, 1954; Harley and Loughman, 1963). It probably first enters a small pool of inorganic phosphate, perhaps within the fungal cytoplasm, from which it is accumulated into the sheath cells in a large pool of inorganic phosphate, together with some organic phosphate. From the small pool phosphate also passes directly to the host tissues. Phosphate accumulated in the large pool in sheath tissue is only released to the host when conditions of low phosphate availability occur around the root (Harley and Brierley, 1954, 1955). Polyphosphate granules, presumably a means of storing large amounts of phosphate, have been detected in the sheath and cortical hyphae of *Pinus* mycorrhiza (Ling-Lee, Chilvers and Ashford, 1975). Whether all ectomycorrhizas resemble that of *Fagus* with respect to the mechanism of phosphate absorption is not yet clear but indications are that differences do exist, some due to physical factors, for example the size of the mycorrhizal organs, others due to as yet undefined physiological factors.

It has frequently been suggested that ectomycorrhizal fungi are able to mobilize soil minerals that are normally insoluble, and in particular sources of phosphate. Analysis of soil around ectomycorrhizas has failed to demonstrate increases in either available phosphorus or in ammonia and potassium, but such negative results may be due to the rapid absorption of any mobilized minerals by the mycorrhizal roots (Mitchell, Finn and Rosendahl, 1937; Stone, 1950). Recent studies have been concentrated on the ability of ectomycorrhizal fungi in axenic culture and ectomycorrhizas to utilize phytates (inositol hexaphosphates). These may constitute the bulk of insoluble phosphates in the organic layers of some soils, and their possible mobilization through hydrolysis by ectomycorrhizas could have obvious and important consequences by releasing inorganic phosphate. At least some ectomycorrhizal fungi can obtain phosphate from calcium and sodium inositol hexaphosphates by means of phytase action (Theodorou, 1968, 1971). It has also been demonstrated that excised beech mycorrhizas possess the enzymes necessary to hydrolyse a number of organic phosphates including p-nitrophenyl phosphate, β-glycerophosphate and inositol hexaphosphate (Bartlett and Lewis, 1973). The activity of p-nitrophenyl phosphatase in beech mycorrhizas in 2—8 times that found in uninfected roots, and histochemical studies suggest that at least some phosphatases are located in the sheath. There is, however, the possibility that they might be present in host tissues, the histochemical reactions necessary for their demonstration being masked by the presence of tannins in the cortical cells (Williamson and Alexander, 1975). Despite the fact that it has been conclusively demonstrated that ectomycorrhizas produce these phosphatases, their significance in mycorrhizal physiology is uncertain. Saprotrophic fungi, bacteria and non-mycorrhizal roots of higher plants have all been shown to be able to use phytate as a phosphorus source, so that this property is not exclusive to the mycorrhizal condition. Although phosphatase activity in mycorrhizas may be higher than that of uninfected roots, what proportion of total phosphate passing into the mycorrhiza is derived from the mobilization of organic phosphate is not known, and it may well be that it constitutes only a small part, since uptake from soluble phosphate sources is also much higher in mycorrhizas than in uninfected roots. Whether they are only in the sheath or are also present in the host, the phosphatases do not diffuse from the mycorrhiza into the surrounding medium but are bound to sites which are probably located in either the cell walls or plasmalemmas of the mycorrhizal tissues.

While the great majority of published studies on mineral uptake have been concerned with phosphate absorption, something is also known of the nitrogen physiology of ectomycorrhizas. Early claims that mycorrhizas were capable of fixing atmospheric nitrogen are now discounted and it is almost certain that nitrogen fixation is restricted to prokaryotes. Mycorrhizas must, therefore, obtain all their nitrogen directly from the soil. Excised mycorrhizas of *Fagus* assimilate ammonium ions, but not nitrate ions, and this has led to the suggestion that the uninfected roots of this host are concerned mainly with nitrate uptake while the mycorrhizal roots absorb only ammonium ions (Carrodus, 1966). Since ammonium compounds are the main source of nitrogen in the litter and humus horizons of some soils this possibility seems to be reasonable. However, in this respect the mycorrhiza of *Fagus* may be unusual, if not unique, since many mycorrhizal fungi can utilize nitrate as sole nitrogen source in axenic culture and can produce nitrate reductase (Trappe, 1967; Lundeborg, 1970). The fungi can also use organic nitrogen sources, particularly amino acids, so that mycorrhizas, and hence the host plant, can perhaps take advantage of sources of soil nitrogen which are released from organic matter in the early stages of the decomposition — mineralization cycle for nitrogen.

As is the case with all plant roots, mycorrhizal roots exhibit considerable selectivity during ion uptake, and although studies have been made of the uptake of rubidium, potassium, sodium and sulphate in *Fagus* mycorrhizas little can be said concerning the ecological implications of the results (Harley and Wilson, 1959; Morrison, 1962; Harley, 1969). However, there is circumstantial evidence that ectomycorrhizal roots are extremely efficient at obtaining minerals from soils in which non-mycorrhizal roots would be able to grow only poorly. Mycorrhizal *Pinus resinosa* trees have been found to grow well on exposed glacial outwash, their growth rates being comparable with those achieved on well-drained sandy soils. This suggests that both major and minor elements can be efficiently absorbed (Wilde and Iyer, 1962). In more detailed studies it has been found that ectomycorrhizas of *Pinus radiata* take up zinc through both absorption by the sheath and a metabolically-dependent absorption process (Bowen, Skinner and Bevege, 1974). Rates of uptake by the latter mechanism are up to 3 times greater than those of non-mycorrhizal roots when the necessary corrections have been made to take into account the larger size of the mycorrhizal roots. In zinc-deficient soils increased uptake, if accompanied by increased movement to the host, could be of obvious importance.

4. Establishment and Stability

The general physiological consequences of ectomycorrhizal symbioses are quite clear. Mineral uptake by the host, particularly of phosphate, is enhanced and its roots may also receive physical protection from soil-borne necrotrophic fungi, while the mycorrhizal fungus is provided with easily assimilable carbon compounds from host photosynthesis. Association leads to ecological success for both partners, allowing them to exploit habitats in which neither alone would thrive. The symbiosis must, however, be first firmly established and second maintained for a long enough period for mutual benefits to accrue.

Infection of virgin roots takes place from the soil, in which potential mycorrhizal fungi lie in the form of resting mycelium, hyphal strands or rhizomorphs, spores or other kinds of propagule. Establishment of a mycorrhiza involves three distinct phases 'recognition' by the fungus of a suitable host, invasion of the root tissues of that host, and, finally, the production of an external sheath. A prerequisite for infection is growth, perhaps only limited growth, of the mycorrhizal fungus within the rhizosphere of its host. A multitude of physical, chemical and microbial factors influence fungal behaviour in the rhizosphere and on the rhizoplane, and it is by no means clear which of these are important in nature in determining successful mycorrhizal infections. Of all the possible factors, it is probable that root exudates play a major role through providing nutrients for fungal growth on or close to the root surface, and specific organic compounds which may enable a mycorrhizal fungus to distinguish between a potential host and a non-host species (Melin, 1963). With respect to the latter compounds, it should be noted that none has yet been positively identified and that colonization of non-host roots does occur in some situations (Theodorou and Bowen, 1971). When hyphal growth of the mycorrhizal fungus in the rhizosphere is initiated the infection process can then begin, but the germinating propagule or young hyphal system must be in a suitable position on the root, since a mycorrhiza can be formed only through invasion of living primary cortical tissue.

In *Pinus radiata* suitable susceptible tissues occur in a wide or narrow zone behind the root apex and in advance of that region where maturation of the primary cortex is occurring (Marks and Foster, 1973). Intercellular invasion of the cortex takes place with little apparent effect on host cells. The walls of the latter separate at the middle lamella and hyphae grow into the gaps so created. An increase in hyphal diameter then forces the host cells further apart (Foster and Marks, 1966, 1967). The sheath forms through anastomosis and compaction of the external weft of hyphae as this overgrows the infected root. Why a compact, frequently smooth, sheath should develop is not clear but it may be due to a combination of a thigmotropic response and an affinity for host root exudates. There is, though, no evidence that root compounds which stimulate growth of ectomycorrhizal fungi can also induce them to form a typical sheath-like structure (Melin, 1963).

The mycorrhizal root system and its sheath effectively constitute a dual organism in much the same way that a fungus and alga combine to form a lichen thallus. In the case of the ectomycorrhiza it is the autotroph that provides the bulk of the somatic tissue of the organism but, as with lichens, the symbiosis results in the formation of an entity with a new, distinct morphology, and which has special functional characteristics. Again, as with lichens, the maintenance of physical and physiological stability under a range of environmental conditions is of great ecological importance for the symbiosis. A large number of investigations have been carried out on the effects of environmental factors on ectomycorrhiza formation, but for the most part no distinction has been made between the operation of these during the infection stage and their effects during the subsequent phases of mycorrhiza establishment. Despite this, it is possible to draw certain tentative conclusions from the results of such studies that relate directly to the nature of those physiological processes which affect the stability of ectomycorrhizas.

In general, although there are exceptions, the mycorrhizal condition is favoured by either high light intensity or a relatively long photoperiod (Björkmann, 1942; Harley and Waid, 1955; Shemakhanova, 1962). This, together with the observation that many tree seedlings become mycorrhizal only upon development of the primary leaves, indicates that photosynthesis, and hence carbohydrate status, may be important for formation and maintenance of ecto-mycorrhizas. This view is supported by the demonstration that explants of *Pinus sylvestris* roots become mycorrhizal if supplied with sucrose through their attached hypocotyls (Fortin, 1966). Availability of carbohydrate to the mycorrhizal fungus is of obvious importance for continuing effective symbiosis but it may also affect the fungus in another way. It has already been mentioned that some strains of some mycorrhizal species produce exo-enzymes that can break down cellulose, hemicellulose and pectic compounds *in vitro,* and it is entirely possible that they are also potentially capable of secreting these enzymes while associated with their host root. It has been suggested that they fail to do so to any extent because high sugar levels in the root tissues repress synthesis of degradative enzymes. While being normally biotrophic they might thus have a capacity to become necrotrophic within roots that have low endogenous sugar levels (Norkrans, 1950; Lewis, 1973). As well as affecting carbohydrate status, light may also influence mycorrhizas through metabolic processes that are unrelated to photosynthesis. This possibility is discussed later.

The role of auxins in the formation and functioning of ectomycor-rhizas has attracted some attention, hyperauxiny being a characteristic of mycorrhizal roots. The common features of all mycorrhizal short roots, relative shortness and thickness, lack of root hairs and radial elongation of the cortical cells, can all be reproduced in aseptic roots by application of auxins. It is at present considered that these character-istics are not simply structural abnormalities, but that they reflect specific physiological and metabolic properties possessed by mycorrhizal roots. Although many effects of auxins within mycorrhizal roots await elucidation others, for example induction of translocation into the roots and enhancement of starch hydrolysis in the cortical cells, are of obvious importance for the symbiosis (Slankis, 1961). The origin of hyperauxiny remains to be conclusively determined. Ecto-mycorrhizal fungi produce IAA and related compounds in axenic culture, particularly if they are supplied with tryptophane, and culture filtrates will induce mycorrhiza-like morphological changes in aseptic roots. If mycorrhizal roots for some reason lose their fungus then their characteristic structural features disappear and they come to resemble normal non-mycorrhizal roots (Slankis, 1967). This suggests that fungal synthesis of IAA leads to hyperauxiny of roots. However, it has been argued that, even if the fungi when in the symbiotic state synthesized growth substances at the same rate as that found in axenic culture, then the amount produced would still be too small to induce the characteristic morphological changes. Doubt has also been expressed concerning the host's ability to supply sufficient tryptophane to the fungus for the rate of syntheis of IAA to approach that found in culture.

Ectomycorrhizal fungi in axenic culture also produce substances that are strongly inhibitory to the auxin oxidase systems in host root tissue, and liberation of these by the fungi when in the symbiotic condition could lead to an increase in the plant's own endogenous auxins and hence result in hyperauxiny (Ritter, 1968). Whatever the actual cause of hyperauxiny, it must be accepted that stability and normal functioning of ectomycorrhizas probably depend on the maintenance of a constant supply of IAA to host root tissues, and that deviation from a critical level of growth substances will lead to a loss of stability (Slankis, 1973). Two factors, light intensity and nitrogen availability, affect mycorrhizal stability by virtue of their influence on auxin levels (Slankis, 1971). At low light intensities application of exogenous IAA to aseptic root systems of *Pinus strobus* fails to induce the usual morphogentic response. The reasons for this are not clear, but the effect is independent of the carbohydrate status of the roots (Slankis, 1963). If this situation obtains in mycorrhizal roots then a breakdown of the symbiosis is to be expected at low light intensities.

It has frequently been observed that increased nitrogen availability prevents the establishment of ectomycorrhizas or causes already formed mycorrhizal roots to revert to the non-mycorrhizal condition. This has been considered to be due to the assimilation of nitrogen into organic nitrogen compounds, especially proteins, within the root causing a decrease in endogenous sugar levels (Björkman, 1942, 1944, 1970). This may be so in some instances, but it is apparent that, frequently, the effect of increased nitrogen availability is independent of any changes in carbohydrate status that might occur (Slankis, 1973). There are two possible ways in which excess nitrogen might influence ectomycorrhizal stability. First, an increase in endogenous nitrogen within roots could cause auxins to be converted to closely related compounds which have no growth-promoting activity. Second, high nitrogen conditions may directly inhibit auxin production by the fungus or by the fungally modified host cells. Aseptic roots of *Pinus strobus* respond to exogenous IAA whether or not excess nitrogen is provided to them, and this has been taken to indicate that it is IAA production by the fungus, not the host, which is affected by the supply of nitrogen to mycorrhizal roots (Slankis, 1971). In addition, production of extracellular IAA by mycorrhizal fungi in axenic culture decreases with an increasing supply of readily available nitrogen compounds (Moser, 1959). The mode of action of high nitrogen levels on the symbiosis is far from being fully understood, but in *Pinus strobus* increases in exogenous nitrogen concentration increase the amount of amino acids within ectomycorrhizal roots, and also alter the qualitative amino acid content (Lister, Slankis, Krotkov and Nelson, 1968; Slankis, 1971). Specific patterns of amino acid synthesis seems to be associated with the ectomycorrhizal condition in *Pinus nigra* and *Pinus sylvestris* and, since the presence of exogenous amino acids inhibits IAA synthesis in axenic cultures of mycorrhizal fungi, the mechanism of high nitrogen suppression of mycorrhizas might be linked to amino acid metabolism (Moser, 1959; Krupa, Fontana and Palenzona, 1973; Krupa and Branstrom, 1974). In addition to the effects of auxins on stability of the symbiosis, other substances, for example cytokinins, have been implicated and may have an important but as yet undetermined role (Slankis, 1973).

One further factor is worthy of consideration. *Boletus* (*Suillus*) *variegatus* forms a sheath around hollow silicone rubber 'roots' in tubes of agar if these 'roots' are supplied internally with oxygen which can diffuse out into the surrounding medium, and if the medium contains nutrients. No sheath is formed if the 'roots' are sealed or if nutrients are omitted from the agar (Read and Armstrong, 1972). Oxygen, in conjunction with host nutrients, may therefore be of importance for sheath formation and persistence. In the humus layers of soil there is frequently a lack of oxygen, but a diffusion pathway for oxygen exists from the shoot to the root and this may facilitate maintenance of normal physiological activity independently of a copious external supply of oxygen.

Bibliography

Bartlett, E. M. and Lewis, D. H., 1973. *Soil Biol. Biochem.*, 5: 249—257.
Björkman, E., 1942. *Symb. bot. upsal.*, 6: 1—191.
Björkman, E., 1944. *Svensk bot. Tidskr.*, 38: 1—14.
Björkman, E., 1970. *Stud. Forest Suec.*, 83: 1—24.
Bowen, G. D., 1968. *Trans. Int. Congr. Soil Sci. 9th*, 1968, Vol. 2, p. 219.
Bowen, G. D., 1973. In G. C. Marks and T. T. Kozlowski (eds.), *Ectomycorrhizae*, p. 151—205, Academic Press, New York.

Bowen, G. D., Skinner, M. F. and Bevege, D. I., 1974. *Soil Biol. Biochem.*, 6: 141—144.

Carrodus, B. B., 1966. *New Phytol.*, 65: 358—371.

Chilvers, G. A. and Pryor, L. D., 1965. *Aust. J. Bot.*, 13: 245—249.

Dominik, T., 1959. *Mycopath. Mycol. appl.*, 11: 359—365.

Fortin, J. A., 1966. *Can. J. Bot.*, 44: 1087—1092.

Foster, R. C. and Marks, G. C., 1966. *Aust. J. biol. Sci.*, 19: 1027—1038.

Foster, R. C. and Marks, G. C., 1967. *Aust. J. biol. Sci.*, 20: 915—926.

Hacskaylo, E., (ed.) 1971. *Mycorrhizae, Proc. 1st N. Am. Conf. Mycorrhizae*, Washington DC.

Hacskaylo, E., 1973. In G. C. Marks and T. T. Kozlowski (eds.), *Ectomycorrhizae*, p. 207—230, Academic Press, New York.

Harley, J. L., 1969. *The Biology of Mycorrhiza*, Leonard Hill, London.

Harley, J. L., 1970. In L. C. Luckwill and C. V. Cutting (eds.), *Physiology of Tree Crops*, p. 163—179, Academic Press, London.

Harley, J. L., 1971. *Mycorrhiza*, Oxford Biology Readers No. 12, London.

Harley, J. L. and Brierley, J. K., 1954. *New Phytol.*, 53: 240—252.

Harley, J. L. and Brierley, J. K., 1955. *New Phytol.*, 54: 296—301.

Harley, J. L., Brierley, J. K. and McCready, C. C., 1954. *New Phytol.*, 53: 92—98.

Harley, J. L. and Jennings, D. H., 1958. *Proc. R. Soc.* B. 148: 403—418.

Harley, J. L. and Loughman, B. C., 1963. *New Phytol.*, 62: 350—359.

Harley, J. L. and McCready, C. C., 1952a. *New Phytol.*, 51: 56—64.

Harley, J. L. and McCready, C. C., 1952b. *New Phytol.*, 51: 342—348.

Harley, J. L. and Waid, J. S., 1955. *Pl. Soil*, 7: 96—112.

Harley, J. L. and Wilson, J. M., 1959. *New Phytol.*, 58: 281—298.

Krupa, S. and Branstrom, G., 1974. *Physiologia Pl.*, 31: 279—283.

Krupa, S., Fontana, A. and Palenzona, M., 1973. *Physiologia Pl.*, 28: 1—6.

Lamb, R. J., 1974. *Trans. Br. mycol. Soc.*, 63: 295—306.

Lewis, D. H., 1973. *Biol. Rev.*, 48: 261—278.

Lewis, D. H. and Harley, J. L., 1965a. *New Phytol.*, 64: 224—237.

Lewis, D. H. and Harley, J. L., 1965b. *New Phytol.*, 64: 238—255.

Lewis, D. H. and Harley, J. L., 1965c. *New Phytol.*, 64: 256—269.

Lindeberg, G., 1948. *Physiologia Pl.*, 1: 196—205.

Ling-Lee, M., Chilvers, G. A. and Ashford, A. E., 1975. *New Phytol.*, 75: 551—554.

Lister, G. R., Slankis, V., Krotkov, G. and Nelson, C. D., 1968. *Ann. Bot.*, 32: 33—43.

Lundeborg, G., 1970. *Stud. Forest Suec.*, 79: 1—95.

Lyr, H., 1963. In W. Rawald and H. Lyr (eds.), *Mykorrhiza*, Fischer, Jena.

McComb, A. L., 1938. *J. For.*, 36: 1148—1154.

McComb, A. L., 1943. *Bull Ia. Exp. Sta.*, 314: 582—612.

McComb, A. L. and Griffith, J. E., 1946. *Pl. Physiol., Lancaster*, 21: 11—17.

Marks, G. C. and Foster, R. C., 1973. In G. C. Marks and T. T. Kozlowski (eds.), *Ectomycorrhizae*, p. 1—41, Academic Press, New York.

Marks, G. C. and Kozlowski, T. T., (eds.) 1973. *Ectomycorrhizae*, Academic Press, New York.

Marx, D. H., 1973. In G. C. Marks and T. T. Kozlowski (eds.), *Ectomycorrhizae*, p. 351—382, Academic Press, New York.

Melin, E., 1963. In P. S. Nutman and B. Mosse (eds.), *Symbiotic Associations*, p. 125—145. 13th. *Symp. Soc. gen. Microbiol.*, Cambridge University Press.

Melin, E. and Nilsson, H., 1950. *Physiologia Pl.*, 3: 88—92.

Melin, E. and Nilsson, H., 1953. *Nature, Lond.*, 171: 134.

Melin, E. and Nilsson, H., 1955. *Svensk bot. Tidskr.*, 49: 119—122.

Melin, E. and Nilsson, H., 1957. *Svensk bot. Tidskr.*, 51: 166—186.

Melin, E. and Nilsson, H., 1958. *Bot. Notiser.*, 111: 251—256.

202

Melin, E., Nilsson, H. and Hacskaylo, E., 1958. *Bot. Gaz.*, 119: 241—246.
Meyer, F. H., 1963. *Ber. dt. bot. Ges*, 76: 90—96.
Meyer, F. H. 1966. In S. M. Henry (ed.), *Symbiosis*, Vol. 1, p. 171—255, Academic Press, New York.
Meyer, F. H., 1974. *A. Rev. Pl. Physiol.*, 25: 567—586.
Mitchell, H. L., Finn, R. F. and Rosendahl, R. O., 1937. *Black Rock For. Pap.*, 1: 58—73.
Morrison, T. M., 1962. *New Phytol.*, 61: 21—27.
Moser, M., 1959. *Arch. Mikrobiol.*, 34: 251—269.
Norkrans, B., 1949. *Svensk bot. Tidskr.*, 43: 485—490.
Norkrans, B. 1950. *Symb. bot. upsal.*, 11: 1—126.
Palmer, J. G. and Hacskaylo, E., 1970. *Physiologia Pl.*, 23: 1187—1197.
Peyronel, B., Fassi, B., Fontana, A. and Trappe, J. M., 1969. *Mycologia*, 61: 410—411.
Read, D. J. and Armstrong, W., 1972. *New Phytol.*, 71: 49—53.
Reid, C. P. P. and Woods, F. W., 1969. Ecology, 50: 179—189.
Ritter, G., 1968. *Acta Mycol. Warsaw*, 4: 421—432.
Shemakhanova, N. M., 1962. *Mycotrophy in Woody Plants, Acad. Nauk. SSSR, Moscow*, (trans., 1st. Programme Sci. Transl. Jerusalem, 1967).
Shiroya, T., Lister, G. R., Slankis, V., Krotkov, G. and Nelson, C. D., 1962. *Can. J. Bot.*, 40: 1123—1126.
Skinner, M. F. and Bowen, G. D., 1974. *Soil Biol. Biochem.*, 6: 53—56.
Slankis, V., 1961. In K. V. Thimann (ed.), *Recent Advances in Botany*, p. 1738—1742, Ronald, New York.
Slankis, V., 1963. In W. Rawald and H. Lyr (eds.), *Mykorrhiza*, p. 175—183, Fischer, Jena.
Slankis, V., 1967. *Proc. 14th. Int. Union Forest Res. Organ. Congr.*, Vol. 5, p. 84.
Slankis, V., 1971. In E. Hacskaylo (ed.), *Mycorrhizae, Proc. 1st. N. Am. Conf. Mycorrhizae*, Washington D.C.
Slankis, V., 1973. In G. C. Marks and T. T. Kozlowski (eds.), *Ectomycorrhizae*, p. 231—298, Academic Press, New York.
Stone, E. L., 1950. *Proc. Soil Sci. Soc. Am.* (1949), 14: 340—345.
Theodorou, C., 1968. *Trans. Int. Congr. Soil Sci. 9th*, Vol. 3, p. 483.
Theodorou, C., 1971. *Soil. Biol. Biochem.*, 3: 89—90.
Theodorou, C. and Bowen, G. D., 1971. *Aust. Forest*, 35: 17—26.
Trappe, J. M., 1962. *Bot. Rev.*, 28: 538—606.
Trappe, J. M., 1967. *Proc. 14th. Int. Union Forest Res. Organ. Congr.*, Vol. 5, p. 46.
Wilde, S. A. and Iyer, J. G., 1962. *Ecology*, 43: 771—774.
Williamson, B. and Alexander, I. J., 1975. *Soil Biol. Biochem.*, 7: 195—198.
Woods, F. W. and Brock, K., 1964. *Ecology*, 45: 886—889.
Zak, B., 1964. *A. Rev. Phytopath.*, 2: 377—392.
Zak, B., 1973. In G. C. Marks and T. T. Kozlowski (eds.), *Ectomycorrhizae*, p. 43—78, Academic Press, New York.

Chapter 12

Endomycorrhizas

It is probably the rule rather than the exception that the roots of plants growing under natural conditions contain a wide range of endophytic fungi. Some of these are antagonists of various kinds, but a proportion are either neutrals or species that are mutualistic to some degree with the plant. Those that are presently considered to be neutral, for example some Chytridiomycete species and filamentous fungi that are looked upon as being harmless endophytes, may prove on further investigation to be either antagonistic or mutualistic, although they have no obvious beneficial or detrimental effects on the host root. The majority of endophytic species may be casual root colonizers in the sense that they are not consistently associated with any particular host. However, it has been shown that the roots of a large number of plant species contain characteristic fungal mycelia which are constantly present when the plants are growing under natural conditions.

It is by no means always clear whether the formation of such associations results in a selective advantage to the host, the fungus or both, and the physiological bases of these symbioses are for the most part still only partially understood (Harley, 1969, 1971; Sanders, Mosse and Tinker, 1975). The majority of endomycorrhizas have two salient characteristics. First, the fungi invade the roots from the soil, which implies that they either have good survival mechanisms that allow them to remain in the soil in a viable condition, or that they have the ability for a free-living saprotrophic existence. Second, when within roots the fungi usually, though not invariably, penetrate host cells, and when this occurs there is minimal host tissue damage. This is an important characteristic of biotrophic fungi but in many endomycorrhizas minimal cell damage appears to be at least partly due to the ability of the host's cytoplasm to digest intracellular hyphae.

Two broad series of endomycorrhizas can be distinguished, one in which the participating fungi are septate and one in which they are consistently aseptate. Mycorrhizas with septate fungi can be further subdivided into those with hosts in the Ericaceae and related families, those with hosts in the Orchidaceae (mentioned in Chapter 8 as non-mutualistic associations), and a miscellaneous group where the hosts are from various and widely scattered plant groups and are not restricted even to distinct phyla (Harley, 1969; Lewis, 1973). The account of endomycorrhizas which follows is mainly restricted to those where there is some experimental evidence that the nature of the fungus—root association is truly mutualistic.

1. Ericaceous Mycorrhizas

The root systems of members of the Ericaceae, and the closely related family the Epacridaceae, normally terminate in a series of very fine, branched, absorbing rootlets. These hair roots consist of a narrow, central stele which is surrounded by one to three layers of cortical cells, there being neither root hairs nor a piliferous layer. The cortical cells of these unsuberized, fine rootlets become consistently invaded by an endomycorrhizal fungus (Burgeff, 1961; McNabb, 1961; Pearson and Read, 1973a).

Features of Infection

The endomycorrhizal fungus forms a more or less prominent weft of septate hyphae growing over the root surface and from these fine, lateral hyphae penetrate the cortical cells, forming well-developed intracellular coils or hyphal masses (Figures 60 and 61). Intracellular hyphae are apparently biotrophic. In actively growing rootlets, the apical meristem and elongating region remain free from infection, and

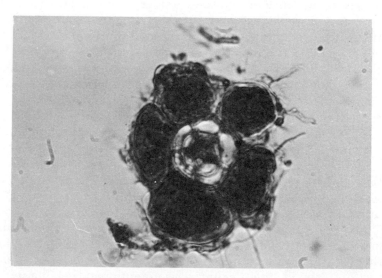

Figure 60   *Calluna vulgaris.* Transverse section of hair root showing cortical cells packed with endophytic mycelium. Photograph by Dr. D. J. Read

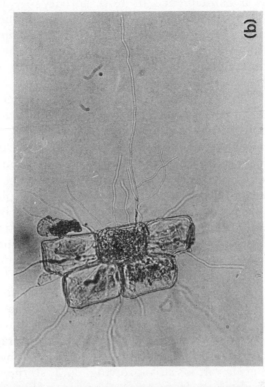

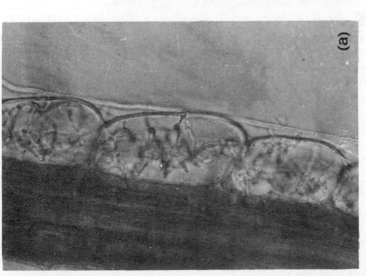

Figure 61 *Calluna vulgaris*: (a) cortical cells of hair root with coiled endophytic hyphae; (b) infected host cells on agar with hyphae emerging from them. Photographs by Dr. D. J. Read

in all roots the stele is never penetrated. After what is apparently a relatively long period of time, the intracellular hyphae become lysed, probably as a consequence of host activity, hyphal digestion being accompanied by an increase in size of host cell nuclei (Rayner, 1927; McNabb, 1961; Nieuwdorp, 1969). It has been claimed, and frequently re-stated, that the typical fungal infection of the Ericaceae extends from the root system to the stem, leaves, flowers and seeds, and that the mycorrhizal fungus is seed-borne, young seedlings becoming infected by means of mycelium growing into the tissues from the seed coat (Rayner, 1927, 1929). It is now clear that such is not the case and that any fungal mycelium occurring in the aerial parts of ericaceous plants, if indeed this is not simply an artifact, is certainly not that of the endomycorrhizal fungus (Harley, 1969; Pearson and Read, 1973a).

Numerous attempts have been made to bring the fungi of ericaceous mycorrhizas into axenic culture, but it is probable that a good proportion of those reported as being successful have not resulted in the isolation of the correct fungus. This is particularly true of the earliest published attempts where a number of strains of *Phoma radicis* were obtained from the roots of species of *Andromeda, Erica* and *Vaccinium* (Ternetz, 1907; Rayner and Levisohn, 1940), and from the seed coat of *Calluna* (Rayner, 1915; Rayner and Smith, 1929). There is no direct evidence that the numerous strains of *P. radicis* are mycorrhizal, and where infection experiments have been carried out these have failed to result in synthesis of the characteristic ericaceous mycorrhiza (Table 16). Typical mycorrhizas have, however, been synthesized using various slow-growing, dark, sterile mycelia that can be consistently obtained from ericaceous roots (Figure 61) (Doak, 1928; Freisleben, 1933, 1934, 1936; Bain, 1937; Burgeff, 1961; McNabb, 1961; Pearson and Read, 1973a). One isolate of the endophyte of *Calluna vulgaris, Mycelium radicis callunae,* has been found to produce a Discomycete perfect state (*Pezizella ericae*) in axenic culture (Read, 1974). It is possible that other isolates from *Calluna,* and those from other ericaceous genera, will ultimately be recognized as being Discomycete species. Isolates of *Mycelium radicis,* with proven ability to form true mycorrhizas with the host from which they were originally obtained, also commonly form typical mycorrhizas with species and genera distinct from their original host (Table 16). (Freisleben, 1934, 1936; Pearson and Read, 1973a). This could either mean that many ericaceous plants share a single mycorrhizal fungus or that a number of distinct fungi with wide host ranges are involved. All the mycorrhizal isolates so far studied have a slow mycelial growth rate but are otherwise extremely variable, particularly with respect to their pigmentation, so that it seems unlikely that a single species is involved.

It has already been mentioned that penetration of the cortical cells of the rootlets is brought about by hyphae arising from mycelium

Table 16

Interspecific synthesis of mycorrhizas in sterile seedlings by fungi isolated from the roots of ericaceous hosts, and by *Phoma radicis callunae*. +, typical mycorrhiza: −, no mycorrhiza (after Pearson and Read, 1973*a*)

Original host of endophyte	Host tested					
	Calluna vulgaris	*Vaccinium myrtillus*	*V. oxycoccos*	*V. macrocarpon*	*Erica cinerea*	*Rhododendron ponticum*
Calluna vulgaris	+	+	+	+	+	+
Vaccinium myrtillus	+	+	+	+	+	+
V. oxycoccos	+	+	+	+	+	+
V. macrocarpon	+	+	+	+	+	+
Erica cinerea	+	+	+	+	+	+
Rhododendron ponticum	+	+	+	+	+	+
Phoma radicis callunae from *Calluna* (isolate of M.C. Rayner)	−	−	−	−	−	−

growing over the rootlet surfaces. Passage of the mycorrhizal fungus between cortical cells, although it does occur, is a comparatively rare event, or is very restricted, cells usually being infected by means of hyphae growing on or near to the rootlet surface (Gordon, 1937; McNabb, 1961). In addition, excised mycorrhizal rootlets are poor sources of inoculum for the mycorrhizal fungus, and in pot-culture of ericaceous plants there is no pot to pot infection even when the containers are in contact (Brook, 1952). Although the evidence is circumstantial, these observations point to the conclusion that the endophyte develops from soil-borne propagules of some kind, and it has been stated that such inoculum is widespread in soils, even those from which an ericaceous flora is normally absent (Gordon, 1937; Brook, 1952). There is some recent evidence to support this contention and it has been established that 75% of dark, slow-growing mycelia isolated directly from soils, including mycelia from some soils with no ericaceous vegetation, are capable of forming mycorrhizas with *Calluna vulgaris* (Pearson and Read, 1973a). However, in what form these mycorrhizal fungi exist in soil is not known but is is entirely possible that, despite their slow rates of mycelial extension, they have some capacity for a free-living saprotrophic existence.

Nitrogen and Phosphorus Physiology

Although some ericaceous plants are calcicoles, many are typically inhabitants of acid soils of low nutrient status. Large areas of the Northern Hemisphere are covered by such soils yet these support good growth of *Calluna, Erica* and *Vaccinium* species, so that is is tempting to assume that the success of ericaceous species in such habitats is a result of the infection of their root systems by mycorrhizal fungi. It has been suggested that the success of ericaceous plants might be due to the endophyte exerting an antagonistic effect on other soil micro-organisms, particularly the mycorrhizal associates of other, and potentially competing, plants (Handley, 1963). It is, however, likely that the major effect of the mycorrhiza is by direct action on host growth and physiology, although few convincing experimental investigations have been made of this possibility. Mycorrhizal infection promotes seedling growth, particularly on nutrient-deficient soils, and in *Calluna vulgaris* and *Vaccinium macrocarpon* the nitrogen and phosphorus content of mycorrhizal seedlings has been found to be significantly higher than that of non-mycorrhizal seedlings (Friesleben, 1933; Bain, 1937; Brook, 1952; Morrison, 1957; Read and Stribley, 1973; Stribley, Read and Hunt, 1975). *V. macrocarpon* seedlings grown on sterile soil for 6 months after artificial infection with their endophyte contain over twice as much nitrogen on a dry weight basis as

Table 17
Nitrogen content and yield of mycorrhizal and non-mycorrhizal seedlings of
Vaccinium macrocarpon. Each figure is the mean from 15 seedlings except at
stage 1 where 12 plants were analysed (from Read and Stribley, 1973)

Growth stage	Total nitrogen content mg/plant	Total dry weight yield per plant (mg)	Appearance
1. Sterile seedlings at time of inoculation	0.11	18.70	
2. Three months after inoculation mycorrhizal	0.88	124.60	Leaves green, plants vigorous
non-mycorrhizal	0.62	121.0	Apex yellow-green, lower leaves reddening
3. Six months after inoculation mycorrhizal	1.82	235.0	Leaves bright-green, plants vigorous
non-mycorrhizal	0.81	184.0	Leaves purple, older even senescent and browning

uninoculated seedlings of the same size. Mycorrhizal seedlings are also larger and more healthy in appearance than their non-mycorrhizal counterparts (Table 17). The basis for enhanced nitrogen content is not altogether clear but it may be that the mycorrhizal roots can absorb soil nitrogen compounds that are for some reason unavailable to non-mycorrhizal roots (Stribley and Read, 1974b). Mycorrhizal seedlings of *V. macrocarpon*, inoculated with an endophyte initially isolated from roots of this species, and grown on ^{15}N-enriched soil (ammonium sulphate) for 6 months show growth stimulation and enhanced nitrogen status. However, uninfected plants grown in the same soil for the same period contain a high proportion of ^{15}N, while infected plants, although they contain greater *total* nitrogen, have a lower ^{15}N-enrichment (Table 18). Mycorrhizal seedlings therefore obviously absorb relatively more unlabelled nitrogen and this implies that they have access to a source of nitrogen, other than the exchangeable ammonium provided, that is unavailable to non-mycorrhizal plants. The source of

Table 18

Nitrogen content, yield and ^{15}N-excess of shoots of mycorrhizal and non-mycorrhizal plants of *Vaccinium macrocarpon* after six months growth in ^{15}N-labelled soil. Each figure is the mean from 14 plants
(from Stribley and Read 1974*b*)

	N content % oven dry wt.	Yield mg oven dry wt.	Total N mg/plant	^{15}N-excess (atom %)
Mycorrhizal	1.20	30.32	0.36	15.38
Non-mycorrhizal	0.98	20.97	0.21	20.03

this nitrogen is not known but there are a number of possibilities. First, mycorrhizal roots might be able to fix atmospheric nitrogen, but this seems to be unlikely. Despite early reports of nitrogen fixation by ericaceous mycorrhizas, current evidence shows that they cannot carry out this process (Ternetz, 1907; Bond and Scott, 1955; Benemann and Valentine, 1972). Second, the mycorrhizal fungus might, through its activities in the soil, release ammonium nitrogen by means of mineralization. However, it has been shown that the *Vaccinium* endophyte does not cause net mineralization in soil (Stribley and Read, 1974*b*). The third, and final, possibility is that the fungus is able to assimilate nitrogen from organic compounds and to conduct this directly into the cortical cells of the rootlets. Free amino acids occur in humus-rich soils that support typical ericaceous vegetation, and these compounds might act as an important nitrogen source for the endophyte. For this to occur the fungus would have to possess an efficient translocatory system to carry compounds from those hyphae lying in the soil to those within the cortical cells. In addition, it would require energy in order to drive those processes involved in nitrogen assimilation and transport. There would also have to be mechanisms acting within infected cells to allow the subsequent release of nitrogenous metabolites from the fungal cells and their absorption by host cells.

There is evidence that a pathway does exist for translocation of nutrients, via the endophyte, from the soil to host roots, and it has been shown that the external mycelium on *Calluna* and *Vaccinium* roots can absorb and translocate phophorus and amino acids into the roots (Pearson and Read, 1973*b*; Stribley and Read, 1974*b*). The mycorrhizal fungus can utilize inositol hexaphosphates as a source of phosphorus, and this might explain the enhanced phosphorus status of mycorrhizal plants grown in nutrient-poor acid soils, where free inorganic phosphorus levels are normally low and the commonest sources of this element are inositol phosphates (Pearson and Read, 1975).

Carbohydrate Physiology

The energy required for nitrogen assimilation and transport is derived ultimately from the carbohydrates within the hyphae of the endophyte. In axenic culture the fungus has a well-developed ability to utilize a number of simple sugars and other, more complex, carbohydrates, for example cellobiose and starch, but has a very restricted capacity to utilize cellulose (Pearson and Read, 1975). In this it resembles some of the fungi of sheathing mycorrhizas. Although the endophyte may have sufficient cellulolytic capability to aid in its penetration of host cell walls, its ability to use cellulose as a major carbon source seems likely to be minimal. Its carbon nutrition may, therefore, be to some degree host-dependent. Hyphae of the endophyte can translocate carbon compounds, and movement of ^{14}C-labelled photosynthates from mycorrhizal seedlings of *Vaccinium* to both the internal hyphae and to the mycelium outside the root cells has been demonstrated (Pearson and Read, 1973*b*; Stribley and Read, 1974*a*). Host photosynthates accumulate within the endophyte in the form of mannitol, trehalose, and also, possibly, a mannose-containing polymer. Movement of carbon from host to fungus is considerable, for instance 72 hours after exposing *Vaccinium* seedlings to labelled carbon dioxide, up to 25% of the total radioactivity detectable in the soluble carbohydrate fraction of mycorrhizal roots may be in the form of fungal carbohydrates. This supply by the host of easily assimilable carbohydrate could be of some ecological advantage to the endophyte but to what degree is uncertain.

It is clear that ericaceous plants benefit from mycorrhizal infections through improved phosphorus and nitrogen nutrition and that the endophyte receives and metabolizes considerable amounts of host photosynthate, upon which it may be at least partially dependent. The mycorrhizal relationship, therefore, appears to be truly mutualistic but there is the possibility, although it is perhaps an unlikely one, that this is not so. Mention has already been made of the eventual lysis of endophytic mycelium by host cells. This may obviously release nutrients to host cells but in addition may reflect, as is the situation in orchidaceous mycorrhizas, a state of necrotrophic antagonism by the plant towards the fungus (Lewis, 1973). Thus, although the fungus receives host photosynthate, at least part of this is subsequently retrieved by the host when the hyphae are digested. If the external hyphae of the mycorrhiza could obtain a significant amount of carbon compounds from their saprotrophic acitivities in the soil then net flow of carbon into the host might occur during symbiosis. In such a situation the advantages to the fungus, if any, of association would be only temporary. In view of the general lack of knowledge concerning the saprotrophic activities of the endophytic fungi of the Ericaceae such a view is entirely speculative.

Some genera in the Ericaceae and members of some other families in the Ericales possess ectendomycorrhizas intermediate in form between the endomycorrhizas typical of the Ericaceae and the ectomycorrhizas of forest trees (Harley, 1969). In the Ericaceae, *Arbutus* and *Arctostaphylos* have such mycorrhizas, and these are characterized by the formation of a fungal sheath around the root, but there is also extensive intracellular penetration by the fungus. The root system of *Arbutus* is differentiated into long and short roots and the latter are invested with a sparse hyphal sheath. These roots assume a swollen, tubercular form and their outer cortical cells are penetrated. Intracellular hyphal coils are produced which eventually become digested and disintegrate. After this has taken place the empty host cells may become recolonized by the fungus and the digestive process is then repeated (Rivett, 1924). Little is known of the physiology of the mycorrhizas of *Arbutus* or *Arctostaphylos*, nor of the fungi that are involved in them. However, polyphosphate granules, similar to those found in the ectomycorrhiza of *Pinus* have been detected in the sheath, intercellular mycelium and intracellular hyphae of *Arbutus* mycorrhiza.

Monotropa

Some members of two other families in the Ericales, the Monotropaceae and the Pyrolaceae, have mycorrhizas resembling in some respects that of *Arbutus*. *Monotropa* is an achlorophyllous, herbaceous genus and from the seedling stage the roots are invested by a compact fungal sheath from which hyphae spread into the surrounding soil, either individually or aggregated into mycelial strands. Hyphae ramify between the cells of the cortex to form a prominent net and also penetrate the outermost cortical cells (Francke, 1934; Meyer, 1966). Each penetrated cell contains a single invasion hypha, all but the tip of which may become encapsulated by a cellulose wall laid down by the host. The invasion hypha grows towards the nucleus of the host cell and when it comes close to this its tip swells and bursts, releasing its contents into the host cell. The mycorrhizal fungus has been brought into axenic culture and has been identified tentatively as being a species of *Boletus*, perhaps *B. subtomentosus* or *B. chrysenteron*, fungi notable for their ability to form sheathing mycorrhizas of forest trees (Furman and Trappe, 1971).

Since *Monotropa* lacks chlorophyll it must obtain its major carbon compounds in an already elaborated form. Laboratory experiments have shown that in the absence of the mycorrhizal fungus seeds of *Monotropa hypopitys* will germinate, but that seedling development will not proceed until a mycorrhiza has become established, and even then growth is slow (Francke, 1934). This suggests that the fungus supplies its host with at least some of the nutrients necessary for normal growth. It has also been shown that the mycorrhizal fungus of *M. hypopitys* can form sheathing mycorrhizas with species of *Pinus* and *Picea*, and it has been established that in nature *M. hypopitys* is linked, through hyphae or mycelial strands arising from its mycorrhiza, to the mycorrhizas of forest trees among which it is growing. If the root systems of vigorously growing plants of *M. hypopitys* are physically separated from the roots of their associated forest trees by means of metal plates driven into the soil, then subsequent growth of *Monotropa* is poor (Björkman, 1960). This demonstrates that the trees are involved in its nutrition and there is conclusive evidence that nutrients, particularly carbon compounds, move from these trees to *Monotropa* via the common mycorrhizal fungus. If ^{14}C-labelled glucose and ^{32}P-labelled orthophosphate are injected into the phloem of mature *Pinus* and *Picea* trees that have *Monotropa* plants growing close to them, then 4—5 days after injection radioactivity can be detected in the *Monotropa* plants, particularly in younger, vigorously growing individuals, up to 2 m away. Any individuals of *Vaccinium* and *Calluna* growing within the same zone do not accumulate radioactivity. Movement of ^{32}P-labelled metabolites has also been shown to take place from *Monotropa* to its associated trees, and radioactivity can be detected in tree root systems 38 cm away from *Monotropa* plants 2 hours after the latter have been injected with ^{32}P-phosphoric acid (Furman, 1966).

The mycorrhizal fungus that is shared between *Monotropa* and trees acts as a nutrient bridge between the heterotrophic plant and its auto-trophic supply of metabolites. The basic requirement of this tripartite symbiosis is obviously the presence of a compatible autotrophic plant which is able to form a mycorrhiza with the same fungus, or fungi, that forms the *Monotropa* mycorrhiza. Nutrients are probably released into the cells of *Monotropa* during the process of hyphal breakdown. This release, as in the endomycorrhizas of the Ericaceae, may be under host control, so that *Monotropa* might be considered to be necrotrophic upon its fungus. This situation has been referred to as 'mycotrophy' (Furman and Trappe, 1971). There remains, however, at least the possibility that the mycorrhizal fungus of *Monotropa* contributes directly, even if only partially or only during the earliest phases of seedling development, to the nutrition of its achlorophyllous host through its saprotrophic activities in the soil.

In view of the intracellular lysis of the fungus, possibly by means of host enzymes, it is arguable that the *Monotropa*—fungus association is not truly mutualistic although that between the same fungus and associated trees obviously is. It is certainly not easy to visualize what advantages might accrue to the fungus through its forming a mycorrhiza with an achlorophyllous host. However, it is not clear what proportion of its nutrients, particularly carbon compounds, the fungal sheath of

Monotropa roots receives directly from the autotrophic host and what proportion it obtains from *Monotropa* by reabsorption after having first acted as a transporter of nutrients to the latter. Certainly the formation of a distinct and characteristic mycorrhizal sheath with the heterotroph strongly suggests that some kind of ecological or physiological benefit is conferred on the fungus by the association, and it may be of some significance in this respect that root extracts of *Monotropa* stimulate both growth of its mycorrhizal fungus and of some *Boletus* species in axenic culture, but not growth of litter-decomposing fungi (Björkman, 1960).

Pyrola contains species lacking functional leaves but also species in which the leaves are well developed. In either case, mycorrhizas appear to be normally present, although details of their formation and function are by no means fully known (Harley, 1969). In leafed species a sheath of variable thickness and density is formed and intercellular hyphae penetrate the epidermal and subepidermal tissues. Intracellular penetration does occur and host cells may die as a result of this.

2. Vesicular-Arbuscular Mycorrhizas

Endomycorrhizas with aseptate hyphae are the most commonly encountered of all the known kinds of mycorrhizas and it is possible that, for the world flora, the majority of mycorrhizal plant species have this kind of infection. Aseptate endophytes are widespread among both cultivated and wild plants, and are not restricted to any particular plant group, being found in bryophytes, pteridophytes (in which mycorrhizas may be formed in non-root tissue), gymnosperms and angiosperms (Harley, 1969). In the tropics and the Southern Hemisphere, where sheathing mycorrhizas are uncommon, they are the characteristic mycorrhizal fungi of many forest trees (Bayliss, 1967). Of these aseptate endophytes, it is those which form vesicular-arbuscular mycorrhizas that are the most common and widespread and their very ubiquity strongly suggests that for their hosts they might have considerable ecological importance.

Features of Infection

The morphology of vesicular-arbuscular endophytes both within and outside host roots is extremely variable and seems to be in no way related to the nature of either the host or the habitat in which the latter is growing. In addition, there is great variability in the extent of infection of any particular host species. On a single root system the occurrence of the mycorrhizal association is often localized and new infections tend to be sporadic. For these reasons the following general account of the structure of vesicular-arbuscular mycorrhizas is of necessity a composite, and by no means all of them have all the characteristics that are described.

Hyphae of the endophyte are inter- and intracellular within the parenchyma of the cortex. The intracellular hyphae either become coiled or, perhaps more typically, become differentiated into densely branched arbuscules (Figure 62). The ultimate branches of these may swell to form small, spherical bodies which are then shed and degenerate. Large, multinucleate, terminal or intercalary oil rich vesicles may be produced on both intracellular and intercellular hyphae (Figure 63b). Starch disappears from those host cells within which arbuscules form, this often taking place prior to penetration, and host nuclei become greatly enlarged or may divide. Their staining reaction also changes and the nuclei of neighbouring but unpenetrated cells may also enlarge (Gerdemann, 1968; Bahadur, 1969). Arbuscules normally quickly disintegrate, probably being digested by the host, and their oily contents are released into the host cell. The host nucleus then returns to its normal size and starch may reappear within the cell. Other intracellular fungal structures, for example hyphal coils and vesicles, are also digested and vesicular-arbuscular mycorrhizas have been divided into 'digestion types' depending on the kind of organ disintegrating within the host cells (Burgeff, 1943; Meyer, 1966). It is, however, doubtful whether such a classification has any real meaning or value since the final result in all cases is identical, being the release of fungal material to the host cell. There is also at present no evidence that large differences in function exist between mycorrhizas that are characterized by the possession or lack of these different fungal organs.

It has been suggested that the arbuscule can be considered to be a haustorium in as much as it is an intracellular terminal branch that is perhaps involved in the interchange of material between host and fungus, and ultrastructural studies indicate a possible special role for the arbuscule (Becking, 1965; Cox and Sanders, 1974; Cox, Sanders, Tinker and Wild, 1975; Kaspari, 1975; Scannerini, Bonfante and Fontana, 1975). The branches of the arbuscule are surrounded by the host cell's plasmalemma. The cytoplasm of occupied cells is rich in plastids, free ribosomes and endoplasmic reticulum, and contains numerous organelles of various kinds (Figure 62). At the point where the hypha which bears the arbuscule penetrates the host cell wall a papilla of host wall material is formed between the host's plasmalemma and the fungal wall. This papilla completely surrounds the penetrating hypha, but is absent from its distal region and from the branches of the arbuscule itself. When individual arbuscules disintegrate their empty cell walls collapse into a mass that remains enclosed by the host's plasmalemma. While transfer of fungal material to the host is possibly brought about by arbuscule disintegration it may also occur through

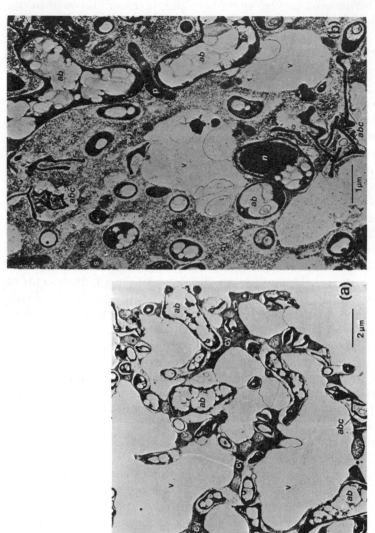

Figure 62   Electronmicrographs of sections through arbuscules of *Glomus mosseae* in *Allium* roots: (a) large and small branches (ab), some of which are collapsed (abc), surrounded by host cytoplasm (cy). The whole arbuscule is located in the vacuole (v) of the host cell; (b) branches embedded in host cytoplasm containing vacuoles. Host cytoplasm is rich in organelles including mitochondria (m), sphaerosomes (s), and plastids (p). From Cox and Sanders, 1974; by permission of *New Phytologist*

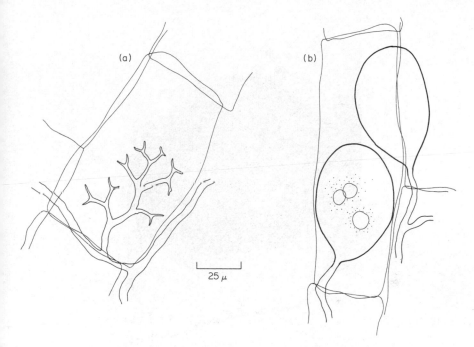

Figure 63   Vesicular-arbuscular mycorrhiza in root of *Allium*: (a) developing arbuscle — part only shown; (b) vesicles

the movement of exocytotic vesicles, although this awaits confirmation (Mosse, 1973). Sometimes groups of arbuscules disintegrate simultaneously and a thick layer of host cell wall material is produced which encases all intracellular structures within that group of host cells. Despite the contention that arbuscules are haustoria their mature structure differs markedly from that of the haustoria of other biotrophic fungi. Although young penetrant hyphae become surrounded by an extrahyphal matrix bounded by a membrane, these being similar to the extrahaustorial matrix and membrane of antagonistic biotrophs, these structures disappear as the arbuscules develop. In addition, in other symbioses where haustoria are a characteristic feature, these organs are usually relatively persistent and stable in structure, yet regular disintegration appears to be the typical situation with respect to the arbuscules of vesicular-arbuscular mycorrhizas. It may be doubted whether they function in a manner similar to that of typical haustoria. Again, it might be noted that vesicular-arbuscular mycorrhizas that lack arbuscules seem to function just as effectively as those that have them.

Hyphae growing within the root are connected to a loose mycelium that extends into the surrounding soil. The hyphae of this mycelium are dimorphic, being composed of persistent, coarse, thick-walled hyphae

Figure 64 Sporocarp of a vesicular-
arbuscular fungus sieved from soil. The
arrow indicates a vegetative spore of the
same fungus. From Mosse and Bowen, 1968;
by permission of *Transactions of the British
Mycological Society*

which have ephemeral, fine thin-walled branches. Thin-walled vesicles,
thick-walled chlamydospores or large sporocarps containing spores may
be borne on the persistent hyphae (Figure 64).

Vesicular-Arbuscular Fungi

Marked differences are found between the numerous vesicular-
arbuscular mycorrhizas that have been described, particularly with
respect to the distribution of the endophyte within the root, the
morphology of the intracellular hyphae, and the possession of vesicles,
chlamydospores and sporocarps. It is, however, apparent that all are
caused by similar kinds of fungi and that these belong to a compact
taxonomic group (Gerdemann, 1968; Mosse, 1973). The fungi are

members of the Endogonaceae, a group which has some affinity with some families in the Mucorales although it is not clear whether it should be included in this order. In the past, nearly all species described from mycorrhizas have been attributed to a single genus *Endogone* but it seems that at least four other genera *Acaulospora, Gigaspora, Glomus* and *Sclerocystis* may be involved and that many fungi formerly ascribed to *Endogone* are species in these genera (Mosse and Bowen, 1968; Gerdemann and Trappe, 1975).

Many attempts have been made to obtain the fungi of vesicular-arbuscular mycorrhizas in axenic culture but with little or no success. Limited hyphal growth on various agar media has been obtained using spores or surface-sterilized mycorrhizal roots as inoculum but subcultures have failed to develop further (Gerdemann, 1968; Mosse, 1973). Successful subculture has been reported, but subsequent use of the same isolation techniques and attempts to infect roots with putative mycorrhizal fungi have in the main been unsuccessful (Barrett, 1961; Mosse, 1961, 1963; Gerdemann, 1968). Successful dual cultures in root organ culture have, however, been obtained (Mosse, 1962; Mosse and Hepper, 1975).

Although axenic cultures of vesicular-arbuscular fungi are lacking, it is possible to carry out experiments on the effects of infection on plant growth by using either infected root fragments or spores as inoculum. The host range of most endophytes appears to be very wide, and material from a single clone is capable of producing infection in hosts from a wide variety of plant groups (Gerdemann, 1961, 1965). It should, however, be noted that there is no guarantee that the inoculum used in most infection experiments was genetically uniform, and there is some evidence that, through employing more meticulous techniques, it is possible to demonstrate some degree of host specificity. Using segments of surface-sterilized roots it has been shown that, while the mycorrhizal fungi of wheat and rye are interchangeable with those of other cereals, those from oats or barley cannot be transmitted either to or from other cereals (Tolle, 1958).

Effects on Host Growth

A large volume of published work exists which is concerned with the effects of vesicular-arbuscular fungi upon the growth of their hosts. Investigations have involved the use of widely different host plants grown under a wide range of soil conditions, and various methods of inoculation have been employed. It is, therefore, not surprising that a broad spectrum of responses have been recorded, but is is also nevertheless quite clear that the most important consequence of infection is enhanced host growth (Gerdemann, 1968; Harley, 1969; Mosse, 1973). This effect is most marked when hosts are growing in phosphorus-deficient soils and it has been shown to be related to increased phosphorus uptake by mycorrhizal plants (Bayliss, 1959; Gerdemann, 1964; Holevas, 1966; Hayman and Mosse, 1971). Infection may also stimulate nodulation in legumes and other nitrogen-fixing hosts but, although the fungi may be present within the tissues of the nodules, they are not involved directly in either their formation or the process of nitrogen-fixation. Increased nodule production is brought about indirectly through the increase in phosphorus content of the host and enhanced vigour of the host plant (Bayliss, McNabb and Morrison, 1963; Crush, 1974).

Vesicular-arbuscular mycorrhizas permit plants to exploit phosphorus-deficient soils but at the same time the amount of endophytic mycelium within the root is influenced by the phosphorus status of the soil, infection being depressed if the phosphorus content is high (Daft and Nicolson, 1969; Crush, 1973). Such conditions, while being unfavourable for intense mycorrhiza formation, may still permit good growth of the fungus to take place close to the root surface in the rhizosphere. Competition for available soil phosphorus then occurs between the fungus and the host root and, since the affinity of the fungus for this element seems to be greater than that of roots, uptake by the host may be reduced and its growth is subsequently retarded in comparison with fungus-free plants (Crush, 1973).

The amount of enhanced growth shown by mycorrhizal plants grown in phosphorus-deficient soils is very variable and can obviously depend on a number of environmental factors other than soil phosphorus levels. With respect to nutritional factors, increased growth will only take place until the amounts of other nutrients become limiting. However, if levels of these other nutrients are then artificially increased there is a rise in fungal infection and an additional increase in the host's growth response (Bayliss, 1970; Crush, 1973). Variations in host response are probably also related to the degree of root hair development, and this in turn varies with the species. Plants which naturally have few or short root hairs are sometimes more dependent on infection by vesicular-arbuscular fungi for growth on phosphorus-deficient soils than are those with good root hair development, although the reverse can also be true (Bayliss, 1972; Mosse, Hayman and Arnold, 1973). The reasons why high soil phosphorus levels can adversely affect mycorrhizal development are not clear, but it is possible that an increased phosphorus content of the host changes either its metabolic patterns or structural features in such a way as to make it resistant to fungal penetration. There is little evidence that phosphorus has a direct and deleterious effect on the mycorrhizal fungi (Mosse and Phillips, 1971; Mosse, 1972).

The possible mechanism for enhanced phosphorus uptake by vesicular-arbuscular mycorrhizas has aroused much interest. The fungus could obviously act as a bridge for phosphorus movement between the soil and infected root cells. Phosphorus compounds would be absorbed and translocated by the external hyphae and then be released within the root, possibly during the process of hyphal dissolution. Given that this is the case, then there are three possible ways in which the fungus might bring about an increased phosphorus content of its host. First, it could increase the movement of phosphate from the labile pool of soil phosphate which contains phosphate normally utilized by the plant. Alternatively, it could mobilize and translocate insoluble or organic forms of phosphorus that are not normally available to the plant. Finally, it could contribute phosphorus compounds from both labile and unavailable sources of phosphate.

It is possible to label the labile phosphate pool in soil while leaving the non-labile sources unlabelled. If a mycorrhizal plant growing in ^{32}P-labelled soil were able to utilize normally unavailable phosphate then, after a suitable experimental period, the activity per unit of phosphate taken up should be less than that found in a plant able to utilize only labile phosphate. However, the specific activity of phosphorus compounds in *Melinus minutiflora*, a tropical grass, when this is grown in ^{32}P-labelled soil is the same irrespective of whether the plants are mycorrhizal or not (Mosse, Hayman and Arnold, 1973). Similarly, the specific activity of phosphorus compounds in mycorrhizal and non-mycorrhizal onions is almost identical, although inflow rates are 3—16 times greater in mycorrhizal plants (Tables 19 and 20), (Sanders and Tinker, 1971; Hayman and Mosse, 1972). All the extra phosphate which enters mycorrhizal plants is therefore coming from the labile pool, and the endophyte is not exploiting a phosphorus source which is normally unavailable to the host.

Inflow rates of phosphorus into mycorrhizal roots, as measured by uptake from labelled solutions, can be considerable (Gray and Gerdemann, 1967; Morrison and English, 1967). Rates are greatly increased if the plant is first starved of phosphate. For example, the

Table 19

Specific activity of phosphate absorbed by mycorrhizal and non-mycorrhizal onion plants of different ages
(after Sanders and Tinker, 1971)

Age of plants (days)	Specific activity cpm x 10^8/mol P	
	mycorrhizal	non-mycorrhizal
23	44	44
37	38	39
44	36	36
54	35	35

Table 20
Mean inflow of phosphorus into mycorrhizal and non-mycorrhizal onion roots
during three successive growth periods
(after Sanders and Tinker, 1971)

Growth period	Duration (days)	Inflow mol/cm root/x 10^{-14}	
		mycorrhizal	non-mycorrhizal
1—2	14	17	5
2—3	7	22	1.6
3—4	10	13	4.2

roots of starved mycorrhizal onion plants take up 160 times as much labelled phosphate as those of non-mycorrhizal plants during a 90 hour feeding period, and their leaves accumulate 45 times as much label as do those of uninfected plants (Gray and Gerdemann, 1969). It is possible that the fungus does not itself take up and translocate phosphate, but that it in some way increases the ability of the roots themselves to do so. However, treating mycorrhizal roots with the fungitoxicant parachloronitrobenzene reduces uptake to the same level as that found in non-myorrhizal roots, and it has also been calculated, using figures from uptake experiments, that an increase in the absorbing ability of the roots alone cannot account for the observed increase in uptake. In addition, translocation of labelled phosphate by vesicular-arbuscular fungi has been demonstrated (Gray and Gerdemann, 1969; Sanders and Tinker, 1971; Hattingh, Gray and Gerdemann, 1973). Autoradiography of labelled sections of mycorrhizal roots shows that labelling is strongest in hyphae, both inside and outside the root, that it is particularly intense in the arbuscules and vesicles, and that it accumulates in polyphosphate granules (Bahadur, 1960; Cox, Sanders, Tinker and Wild, 1975; Ling-Lee, Chilvers and Ashford, 1975). All present evidence therefore points to the conclusion that the external mycelium of vesicular-arbuscular mycorrhizas provides an additional, efficiently distributed surface for absorption. Yet external hyphae are frequently scanty, and even where they are well developed the entry points into the root may still be relatively few in number. However, it has been calculated that neither external hyphal development nor entry points need be abundant in order for significant increases in phosphorus uptake to be brought about (Bieleski, 1973). For instance, 4 hyphal connections per millimetre of root with unbranched hyphae extending 20 mm from the root surface would facilitate a 60-fold increase in phosphorus uptake, if diffusion is limiting, and a 10-fold increase if uptake is proportional to available surface area. The normal extent of the phosphate depletion zone around non-mycorrhizal onion roots is in the region of 1—2 mm. When onion roots are infected with the vesicular-arbuscular endophyte *Glomus fasciculatus* this zone is extended to up to 7 cm from the root surface (Rhodes and Gerdemann, 1975).

Enhanced uptake of nutrients other than phosphorus by vesicular-arbuscular mycorrhizas has been observed and the mycorrhizal condition may alleviate other kinds of mineral deficiencies in nutrient-poor soils. *Glomus mosseae* and *G. macrocarpus* can also reduce nitrate to nitrite (Gilmore, 1971; Jackson, Miller and Franklin, 1973; Ho and Trappe, 1975).

The advantages of fungal infection to the host are quite apparent but, although the host—fungus relationship is almost certainly truly mutualistic, it is not yet clear to what degree the fungus benefits from the association. Difficulties encountered in attempts to obtain cultures of vesicular-arbuscular fungi suggest that there is a high degree of obligation for biotrophic association with host roots, and this view is supported by the observation that as infected roots age the fungi die. Whether this obligation is detemined by ecological pressures, or by exacting nutrient requirements, or both, is not known. A major difficulty is that there is almost total ignorance as to the activity of these fungi in soil in nature, but there is evidence that some of them may have at least a partial ability for a free-living existence and hyphae have been seen to spread through the soil from dead organic material which they were apparently exploiting. There is some evidence for the movement of ^{14}C-labelled photosynthates from the host into the mycorrhizal fungus, but the quantitative and qualitative carbohydrate contents of mycorrhizal and non-mycorrhizal roots seem to be similar, and how much of the total carbon requirement of the fungus is met by the host remains to be determined (Ho and Trappe, 1973; Mosse, 1973).

Since the fungi of vesicular-arbuscular mycorrhizas are species of Zygomycotina it is highly likely that their carbohydrate requirements and physiology will differ from those of septate, endomycorrhizal fungi, and from those of ectomycorrhizal species. It may be of some significance in this respect that, like some symbiotic Oomycetes, they do not seem to accumulate polyols, and their growth is stimulated by inositol and glycerol (Mosse, 1972).

Bibliography

Bahadur, A. 1969. *Arch. Mikrobiol.*, 68: 236—245.
Bain, H. R., 1937. *J. agric. Res.*, 55: 811—835.
Barrett, J. T., 1961. In K. V. Thimann (ed.), *Recent Advances in Botany*, p. 1725—1727, Ronald, New York.
Bayliss, G. T. S., 1959. *New Phytol.*, 58: 274—280.

224

Bayliss, G. T. S., 1967. *New Phytol.*, **66**: 231—243.
Bayliss, G. T. S., 1970. *Pl. Soil*, **33**: 713—716.
Bayliss, G. T. S., 1972. *Search*, **3**: 257—263.
Bayliss, G. T. S., McNabb, R. F. R. and Morrison, T. M., 1963. *Trans. Br. mycol. Soc.*, **46**: 378—384.
Becking, J. H., 1965. *Pl. Soil*, **23**: 213—216.
Benemann, J. R. and Valentine, R. C., 1972. *Adv. microbial. Physiol.*, **8**: 59—81.
Bieleski, R. L., 1973. *A. Rev. Pl. Physiol.*, **24**: 225—252.
Björkman, E., 1960. *Physiologia. Pl.*, **13**: 308—327.
Bond, G. and Scott, G. D., 1955. *Ann. Bot.*, **19**: 67—77.
Brook, P. J., 1952. *New Phyt.*, **51**: 388—397.
Burgeff, H., 1943. *Naturwissenschaften*, **31**: 358.
Burgeff, H., 1961. *Mikrobiologie des Hochmores*, Fischer, Stuttgart.
Cox, G. and Sanders, F. E., 1974. *New Phytol.*, **73**: 901—912.
Cox, G., Sanders, F. E., Tinker, P. B. and Wild, J. A., 1975. In F. E. Sanders, B. Mosse and P. B. Tinker (eds.), *Endomycorrhiza*, p. 297—312, Academic Press, London.
Crush, J. R., 1973. *New Phytol.*, **72**: 965—973.
Crush, J. R., 1974. *New Phytol.*, **73**: 743—749.
Daft, M. J. and Nicolson, T. H., 1969. *New Phytol.*, **68**: 945—952.
Doak, K. D., 1928. *Phytopathology*, **18**: 148.
Francke, H. L., 1934. *Flora, Jena*, **129**: 1—5.
Freisleben, R. 1933. *Ber. dt. bot. Ges.*, **51**: 351—356.
Freisleben, R. 1934. *Jb. wiss. Bot.*, **80**: 421—456.
Freisleben, R. 1936. *Jb. wiss. Bot.*, **82**: 413—459.
Furman, T. E., 1966. *Am. J. Bot.*, **53**: 627.
Furman, T. E. and Trappe, J. M., 1971. *Q. Rev. Biol.*, **46**: 219—225.
Gerdemann, J. W., 1961. *Mycologia*, **53**: 254—261.
Gerdemann, J. W., 1964. *Mycologia*, **56**: 342—349.
Gerdemann, J. W., 1965. *Mycologia*, **57**: 562—575.
Gerdemann, J. W., 1968. *A. Rev. Phytopath.*, **6**: 397—418.
Gerdemann, J. W. and Trappe, J. M., 1975. In F. E. Sanders, B. Mosse and P. B. Tinker (eds.), *Endomycorrhiza*, p. 35—51, Academic Press, London.
Gilmore, A. E., 1971. *J. Am. Soc. Hort. Sci.*, **96**: 35—38.
Gordon, H. D., 1937. *Ann. Bot.*, **1**: 593—613.
Gray, L. E. and Gerdemann, J. W., 1967. *Nature, Lond.*, **213**: 106—107.
Gray, L. E. and Gerdemann, J. W., 1969. *Pl. Soil*, **30**: 415—422.
Handley, W. R. C., 1963. *Forestry Commission Bulletin* No. 36, HMSO, London.
Harley, J. L., 1969. *The Biology of Mycorrhiza*, Leonard Hill, London.
Harley, J. L., 1971. *Mycorrhiza*, Oxford Biology Readers No. 12, London.
Hattingh, M. J., Gray, L. E. and Gerdemann, J. W., 1973. *Soil Sci.*, **116**: 383—387.
Hayman, D. S. and Mosse, B., 1971. *New Phytol.*, **70**: 19—27.
Hayman, D. S. and Mosse, B., 1972. *New Phytol.*, **71**: 41—47.
Ho, I. and Trappe, J. M., 1973. *Nature (New Biol.), Lond.*, **244**: 30—31.
Ho, I. and Trappe, J. M., 1975. *Mycologia*, **67**: 886—888.
Holevas, C. D., 1966. *J. hort. Sci.*, **41**: 57—64.
Jackson, N. E., Miller, R. H. and Franklin, R. E., 1973. *Soil Biol. Biochem.*, **5**: 205—212.
Kaspari, H. 1975. In F. E. Sanders, B. Mosse and P. B. Tinker (eds.), *Endomycorrhiza*, p. 325—334, Academic Press, London.
Lewis, D. H., 1973. *Biol. Rev.*, **48**: 261—278.
Ling-Lee, M., Chilvers, G. A. and Ashford, A. E., 1975. *New Phytol.*, **75**: 551—554.
McNabb, R. F. R., 1961. *Am. J. Bot.*, **9**: 57—61.

Meyer, F. H., 1966. In S. M. Henry (ed.), *Symbiosis*, Vol. 1, p. 171—255, Academic Press, New York.

Morrison, T. M., 1957. *New Phytol.*, 56: 247—257.

Morrison, T. M. and English, D. A., 1967. *New Phytol.*, 66: 243—250.

Mosse, B., 1961. In K. V. Thimann (ed.), *Recent Advances in Botany*, p. 1728—1732, Ronald, New York.

Mosse, B., 1962. *J. gen. Microbiol.*, 27: 509—520.

Mosse, B., 1963. In P. S. Nutman and B. Mosse (eds.), *Symbiotic Associations*, p. 146—170. 13th. *Symp. Soc. gen. Microbiol.* Cambridge University Press.

Mosse, B., 1972. *Rep. Rothamsted Exp. Sta.*, (1971) p. 93.

Mosse, B., 1973. *A. Rev. Phytopath.*, 11: 171—196.

Mosse, B. and Bowen, G. D., 1968. *Trans. Br. mycol. Soc.*, 51: 469—483.

Mosse, B., Hayman, D. S. and Arnold, D. J., 1973. *New Phytol.*, 72: 809—815.

Mosse, B. and Hepper, C., 1975. *Physiol. Pl. Path.*, 5: 215—223.

Mosse, B. and Phillips, J. M., 1971. *J. gen. Microbiol.*, 69: 157—166.

Nieuwdorp, P. J., 1969. *Acta bot. neerl.*, 18: 180.

Pearson, V. and Read, D. J., 1973a. *New Phytol.*, 72: 371—379.

Pearson, V. and Read, D. J., 1973b. *New Phytol.*, 72: 1325—1331.

Pearson, V. and Read, D. J., 1975. *Trans. Br. mycol. Soc.*, 64: 1—7.

Rayner, M. C., 1915. *Br. J. Exp. Biol.*, 2: 265—291.

Rayner, M. C., 1927. *Mycorrhiza, New Phytol.*, reprint 15.

Rayner, M. C., 1929. *Ann. Bot.*, 43: 55—70.

Rayner, M. C. and Levisohn, I., 1940. *Nature, Lond.*, 143: 461.

Rayner, M. C. and Smith, M. L., 1929. *New Phytol.*, 28: 261—290.

Read, D. J., 1974. *Trans. Br. mycol. Soc.*, 63: 381—382.

Read, D. J. and Stribley, D. P., 1973. *Nature (New Biol.) Lond.*, 244: 81—82.

Rhodes, L. H. and Gerdemann, J. W., 1975. *New Phytol.*, 75: 555—561.

Rivett, M., 1924. *Ann. Bot.*, 38: 661—677.

Sanders, F. E., Mosse, B. and Tinker, P. B., (eds.) 1975. *Endomycorrhiza*, Academic Press, London.

Sanders, F. E. and Tinker, P. B., 1971. *Nature, Lond.*, 233: 278.

Scannerini, S., Bonfante, P. F. and Fontana, A., 1975. In F. E. Sanders, B. Mosse and P. B. Tinker (eds.), *Endomycorrhiza*, p. 313—324, Acadèmic Press, London.

Stribley, D. P. and Read, D. J., 1974a. *New Phytol.*, 73: 731—741.

Stribley, D. P. and Read, D. J., 1974b. *New Phytol.*, 73: 1149—1155.

Stribley, D. P., Read, D. J. and Hunt, R., 1975. *New Phytol.*, 75: 119—130.

Ternetz, C., 1907. *Jb. wiss. Bot.*, 44: 353—408.

Tolle, R., 1958. *Arch. Mikrobiol.*, 30: 285—303.

Chapter 13

Lichens

Associations between fungi and algae are common and widespread, and the nature of these associations can vary widely (Ahmadjian, 1967; Hale, 1974). At their simplest there is merely a transient antagonistic or mutualistic interaction between a free-living fungus and a free-living alga. At their most complex, association results in the formation of long-lived, morphologically distinct entities within which there is physical and physiological integration of mutual benefit to both organisms. In such 'dual organisms' the fungus frequently provides the bulk of the total somatic material and there is normally a degree of ecological obligation on the part of one or both partners for association. It is to these dual structures that the term 'lichen' should be strictly applied and it is these which will mainly be discussed here.

It should, however, be pointed out that there are a number of fungus—alga associations which are lichen-like in that they too are long-lived, and within them there is also intimate contact between the partners. In these, the alga normally contributes the bulk of the somatic material and determines the morphology of the dual organism. One of the best-known examples is the association between the marine Ascomycotina *Mycosphaerella ascophylli* and the brown algae *Ascophyllum nodosum* and *Pelvetia canaliculata* (Sutherland, 1915; Weber, 1967). Plants are never found that lack intercellular hyphae, and the association seems to be initiated in the early stages of the life of the alga. Algicolous fungi of this kind appear to be widespread in the marine environment and the term 'mycophycobioses' has been proposed for such fungus—alga associations (Kohlmeyer, 1968; Kohlmeyer and Kohlmeyer, 1972). Whether the presence of the fungus affects the plant in any way is not known, but it is almost certain that mycophycobioses do not function in the same way as lichens (Kremer, 1973). Mycophycobioses have their parallel in the terrestrial environment in forms where filamentous algae become enveloped by fungal hyphae to produce a permanent dual structure. The alga is the predominant partner and, although the fungus undergoes no morphological change, together they form a distinct morphological entity (Karling, 1934). Other lichen-like associations involve some fungi that have a well-developed capacity for a saprotrophic existence but which also have a facultative ability to form unions with algae. In addition, fungi that inhabit the surface of lichens may become associated with the algae within the thalli. Little or nothing is known of the physiology of these symbioses.

Each lichen comprises a single species of fungus, the mycobiont, and usually a single algal species, the phycobiont, although in some lichens a second phycobiont may also be present. The mycobiont is normally a species of Ascomycotina or, less commonly, a Hyphomycete or a species of Basidiomycotina. The phycobiont is almost invariably a green or blue-green alga, *Trebouxia* being the commonest green alga and *Nostoc* or *Scytonema* being the most common blue-green algae (Ahmadjian, 1967). The range of morphology of lichen thalli and their various reproductive structures is very great but, whatever the gross form of the thallus, its algae are usually completely surrounded by fungal tissue and are normally confined to a layer close to the thallus surface. In some lichens this heteromerous arrangement is lacking, and in these types, homoiomerous lichens, the algae are distributed throughout the thallus so that no distinct photosynthetic layer is formed. A constant character of some lichens is the presence of cephalodia. These are small surface outgrowths, or internal organs, containing a second phycobiont, invariably a blue-green species. There is no evidence that every species of lichen contains a phycobiont species exclusive to itself and, on the contrary, it has been shown that many different lichens share the same phycobiont.

It has long been assumed that metabolites produced by the alga are taken up and utilized by the fungus and that the nutrition of the latter is in the main biotrophic. Experimental investigations have confirmed these assumptions but have also demonstrated that fungus—alga relationships' within the thallus are by no means simple. Studies have taken place in two broad areas, one being concerned with the physiological bases of the symbiosis and the other with the effects of environmental conditions on lichens or lichen components. Elucidation of the physiology of the fungus—alga association has been attempted using three major approaches; studies on axenic cultures of separated mycobionts and phycobionts, with attempted artificial synthesis of whole thalli; examination of the ultrastructure of fungal hypha—algal cell contact; and the study of movement of specific metabolites within thallus tissues.

1. Culture of Symbionts and Lichen Synthesis

In axenic culture many, but by no means all, lichen fungi grow relatively slowly, and if spores are used as initial inoculum these may require a period of several weeks for germination to take place (Ahmadjian, 1961; Fox, 1966). With slow-growing species, not only is the rate of growth slow but also maximum mycelial yield obtained under controlled conditions is low. Lichen fungi have a slightly psychrophilic tendency, growing best at 15—20°C, and this may be below the optimum temperature for growth of their particular algal partner (Henriksson, 1964). They resemble most free-living saprotrophic fungi in that they can utilize a wide range of carbon compounds as sole sources of carbon. Some can utilize nitrate nitrogen, although growth rates are increased if the nitrogen source is an amino acid (Ahmadjian, 1964; Gross and Ahmadjian, 1966). There is a general requirement for biotin and thiamine but at least one lichen fungus is autotrophic for the former vitamin. It does, however, require non-vitamin growth factors normally present in yeast extract (Hale, 1958; Richardson and Smith, 1968b). Their frequently reduced growth rates, and their probable inability to compete with other micro-organisms when in the non-lichenized condition, suggests that they are ecologically obligate symbionts. Their taxonomic relationships with their free-living counterparts are frequently obscure since fruiting structures are not invariably formed in nature, and only rarely in axenic culture.

Lichen algae grown axenically have the same autotrophic and heterotrophic nutritional requirements as corresponding free-living species but they tend to grow slowly, having generation times varying from days to weeks (Ahmadjian, 1967; Fox, 1967). Many lichen algae may be ecologically obligate symbionts and Trebouxia species, for example, are rarely found free-living in nature. However, since they are obviously autotrophic, this probably means that at least some of them are capable of a free-living but perhaps not very successful existence. In axenic culture lichen algae release to the medium numerous metabolites, both energy sources and vitamins, of the kind which axenically-grown mycobionts require for growth and, in addition, blue-green algae can fix atmospheric nitrogen (Henriksson, 1951; 1961; Bednar and Holm-Hanson, 1964).

A number of attempts have been made to synthesize lichen thalli from the appropriate axenically-grown symbionts, (Ahmadjian, 1959a, 1962, 1966; Ahmadjian and Heikkilä, 1970). In general, lichenization does not take place under conditions that permit independent growth of both symbionts, but will occur when they are subjected to stress resulting from lack of either nutrients or water. Additionally, the mycobiont must be able to develop sufficiently to allow it to contain algal cells within its mycelium. Most successful syntheses result in the formation of small lichenized structures, but these have such slow growth rates that recognizable thalli have so far not been observed to differentiate. It is doubtful whether lichen synthesis studies can contribute much to the knowledge of lichen physiology, since other techniques are available for the successful examination of inter-symbiont movement of metabolites within intact lichen tissues under controlled conditions.

2. Ultrastructure

Algal cells within thalli undergo the normal cycle of division, growth, maturation, senescence and degeneration, with a positive balance between loss of old cells and production of young ones to replace them. In thalli, algal growth rates are slower and generation times much longer than is found in free-living forms. Within a single thallus the distribution of algal cells of different ages is not uniform and young and mature cells predominate in those areas where thallus growth is taking place.

The physical relationship between mycobiont and phycobiont has been studied in a number of lichens, although in perhaps too few for firm generalizations to be made. At the cellular level there are three kinds of interaction between fungus and alga. Commonly, there is a simple wall to wall contact with very close attachment between the hypha and algal cell, and there may be fusion of the two walls, those of both fungus and alga becoming thinner over the area of contact (Durrell, 1967; Peat, 1968; Walker, 1968; Ben-Shaul, Paran and Galun, 1969; Galun, Paran and Ben-Shaul, 1970b). In the second kind of interaction the hypha grows into the algal cell and the wall of the latter, while it does not rupture, becomes deeply invaginated (Figure 65a). Within the invagination there may be fusion between the two walls, and changes in algal cell wall density have been observed which indicate that fungal enzymes are acting upon it (Durrell, 1967; Chervin, Baker and Hohl, 1968; Galun, Paran and Ben-Shaul, 1970a; Paran, Ben-Shaul and Galun, 1971). Finally, in the third type of interaction, hyphae enter algal cells by penetrating the cell wall to cause invagination of the algal plasmalemma (Figure 65b). Multiple penetrations of this kind can occur (Moore and McAlear, 1960; Galun, Paran and Ben-Shaul, 1970a). Penetration may be effected by solely physical methods but there is evidence that enzymes are involved and lysosome-like organelles have been found within penetrating hyphae (Webber and Webber, 1970). The penetration hyphae usually assume a clavate shape within the algal cell and their walls are usually thinner than those of their parent hyphae.

Penetrant hyphae are usually described as haustoria, those invaginating the cell wall being called 'intramembraneous haustoria' and those that penetrate it being called 'intracellular haustoria' (Plessl, 1963; Ahmadjian, 1967). These terms are confusing since they do not embody an accurate reference to the nature of the penetration. In addition, a hypha that invaginates the host's cell wall and then fuses

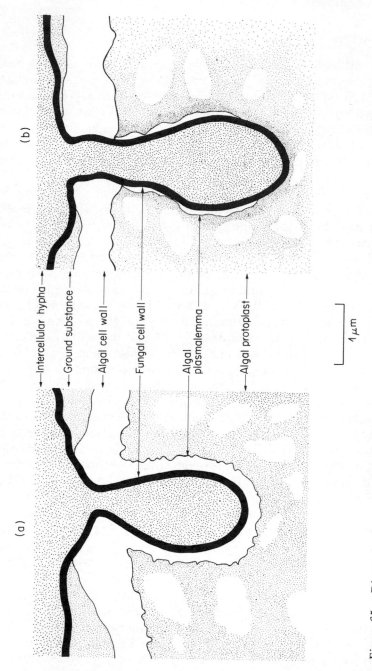

Figure 65   Diagram of mode of penetration of algal cells by hyphae within lichen thalli: (a) invagination but no rupture of host cell wall; (b) rupture of host cell wall and invagination of algal plasmalemma

with it can hardly be referred to as a haustorium. Where rupture of the algal cell wall does occur there is no extrahaustorial matrix, the hyphal wall contacting the host's plasmalemma directly (Roskin, 1970). Structures enclosing the penetrating hyphas have been observed, and variously described as zones of dissolution or spaces, but their nature has not been established (Moore and McAlear, 1960; Durrell, 1967; Galun, Paran and Ben-Shaul, 1970a). Penetration of algal cells by the fungus does not seem to be necessary for the success of the fungus—alga interaction. Many lichens lack penetrant hyphae, and in those lichens that do have them their frequency within the thallus may be very low. In some lichens cell wall invagination or rupture may occur only in post-mature algal cells, when the chloroplast is becoming disorganized, or take place only after death of the cell. There may also be seasonal changes in penetration frequency and, in addition, in different ecotypes of the same lichen species occupying different climatic regions there are wide variations in this frequency (Durrell, 1967; Ben-Shaul, Paran and Galun, 1969; Galun, Paran and Ben-Shaul, 1970a). Although these structures do not appear to be primarily absorptive organs, in those lichens where a significant number of algal cells have their walls penetrated so that the fungus is in contact with the protoplast, it is possible that they could play a part in nutrient transfer to the mycobiont. In doing so they need not necessarily function biotrophically. The frequency with which penetrated host cells die, together with their increasing susceptibility to penetration with advancing age, indicates that the intracellular phase of the fungus is probably characteristically necrotrophic.

It seems that in primitive lichens, where well-defined thalli are lacking, cell wall penetration is much more common than in advanced lichens that have well-defined thalli. There are also transitional types where both penetration and invagination occur within the same thalli (Ahmadjian, 1967; Galun, Paran and Ben-Shaul, 1970a). In primitive thalli all algal cells, irrespective of age, may be penetrated, while in more advanced forms it is only mature and post-mature cells that are normally affected. It also appears, on the limited evidence available, that green algae are more commonly penetrated, or have their walls invaginated, than are blue-green algae (Table 21). The apparent relationship between lack of a well-organized thallus and a high frequency of penetrating hyphae may have some phylogenetic significance, but the situation is open to a number of interpretations. It is tempting to consider the presence of penetrating organs as being a reflection of a lack of balance or compatibility between mycobiont and phycobiont. The disorganized thallus would then be a direct result of a physiologically inefficient symbiosis, or a symbiosis in which antagonism on the part of the mycobiont tended to reduce the benefits which might

232

Table 21
Types of hypha—algal cell interaction in different genera of lichen
algae

| | Type of interaction[a] | | |
	Contact	Invagination	Penetration
Green algae			
Trebouxia	+	+	+
Stichococcus	−	+	+
Myrmecia	−	−	+
Blue-green algae			
Gloeocapsa	+	+	−
Nostoc	+	−	−
Scytonema	−	+	−
Calothrix	+	−	−
Stigonema	+	−	−

[a] + present; − absent.

otherwise accrue to both symbionts from a solely mutualistic interaction. Development by the alga of an ability to resist or tolerate fungal
invasion, together with the evolution of more efficient mechanisms for
metabolite exchange between the symbionts, would result in the
formation of a more organized thallus. Finally, the stage would be
reached where penetration would be absent, the mycobiont being
completely biotrophic and its antagonistic effects reduced to a
minimum. Where an environmental change results in the formation of
penetrating hyphae within thalli from which they are normally absent,
this might be due to the influence of environmental variation on one or
both partners, so upsetting the balance normally obtaining.

The role of penetrating hyphae is enigmatic but other structures are
found in regions of fungus—alga contact to which a function can be
more firmly attributed (Walker, 1968; Peveling, 1973). In the fruticose
lichen *Cornicularia normoerica* penetrating hyphae are absent, but the
algal cells possess channels that originate from the chloroplast and then
extend through the cell wall, flaring out at its periphery (Figure 66a-c).
These channels, which contain small tubules, are relatively infrequent
and occur only in larger algal cells. They may, nevertheless, have some
significance in metabolite exchange and may well occur in other lichen
species. Structural modifications have also been found in the *Calothrix*
phycobiont of *Lichina pygmaea* where the outer layers of the algal cell
wall and the plasmalemma are crenulate (Figure 67a). Vesicles bounded
by a single limiting membrane are present in the fine ground substance
between the algal cell and the fungal hypha, the plasmalemma of the

latter also being invaginated. These vesicles may represent a mechanism of metabolite exchange analogous to that thought to operate in haustoria of some rust fungi, and might be moving from one symbiont to the other through the ground substance to be finally accepted by the cytoplasm of either alga or fungus. Crenulation and invagination of the plasmalemmas of the symbionts could be a device to increase the surface areas of their protoplasts and so increase efficiency of metabolite interchange.

In a number of lichens that have mycobionts in the Ascomycotina, the transverse septa of the hyphae are multiperforate (Figure 67b). In free-living Ascomycotina these septa are usually characterized by a single, simple pore, although in the Hyphomycete genus *Fusarium* multiperforations have been observed. Multiperforations in lichen thalli seem to have little or nothing to do with contact between the symbionts, and perhaps serve mainly to increase efficiency of translocation within the lumina of the hyphae (Wetmore, 1973). Other

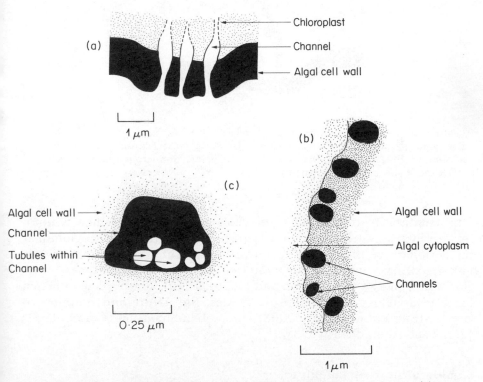

Figure 66   Ultrastructural features of lichens, *Cornicularia normoerica*: (a) channels from within chloroplast extending through algal cell wall; (b) cross-section of channels within cell wall; (c) cross-section of a single channel containing tubules; Diagrams based on a series of electronmicrographs in Walker, 1968

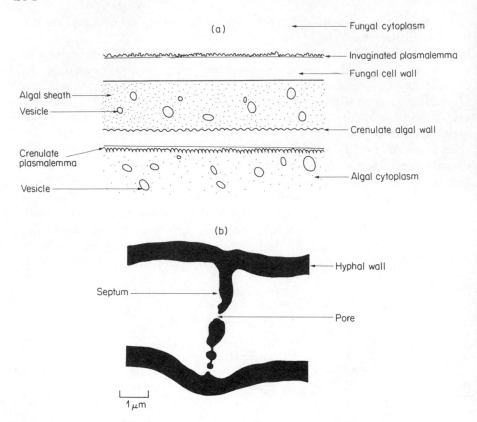

Figure 67　Ultrastructural features of lichens: (a) *Lichina pygmaea*, vesicles in algal cytoplasm and the algal sheath; (b) *Peltigera aphthosa*, multiperforate septum in fungal hypha. Diagrams based on electronmicrographs in Peveling, 1973 and Walker, 1973 respectively

ultrastructural features seem to be characteristic of the lichen symbiosis. For instance, cells of symbiotic green algae contain pyrenoglobuli at the periphery of the pyrenoid and within the chloroplast stroma, while blue-green algae may contain lipid-rich globuli and, sometimes, polyglucoside granules (Ahmadjian and Jacobs, 1970). The hyphae of lichen fungi invariably contain ellipsoid bodies, the nature and function of which are unknown (Peat, 1968; Paran, Ben-Shaul and Galun, 1971; Jacobs and Ahmadjian, 1973).

3. Movement of Metabolites

Studies on metabolite movement have concentrated on two processes; first, the movement of photosynthetic carbohydrate from phycobiont to mycobiont and second, in those lichens containing blue-green algae, the fixation of atmospheric nitrogen and the movement of nitrogen compounds.

Carbohydrate Metabolism

Mention has already been made of the leakage of metabolites from phycobiont cells grown in axenic culture, and of the clear implication that in the thallus these compounds might normally be utilized by the mycobiont. This possibility has been examined in a number of lichens, and leakage of photosynthate from the *Nostoc* symbiont of *Peltigera polydactyla* and the *Trebouxia* symbiont of *Xanthoria aureola* has been studied in some detail (Bednar and Smith, 1966; Drew and Smith, 1967a; Richardson and Smith, 1968a).

Symbiotic *Nostoc* cells, separated from thalli and allowed to photosynthesize using labelled carbon dioxide in a liquid medium, release more than 50% of the total carbon fixed to the medium. The fixed carbon which is not released is incorporated approximately equally into intracellular soluble and insoluble compounds (Figure 68a).

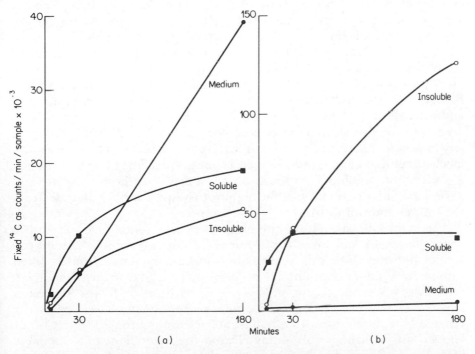

Figure 68   *Peltigera polydactyla*, distribution of fixed ^{14}C between the medium and the soluble and insoluble fractions of *Nostoc* cells. (a) *Nostoc* directly isolated from the thallus; (b) *Nostoc muscorum.* Modified from Drew and Smith, 1967a; by permission *New Phytologist*

Table 22
Release of photosynthetically fixed ^{14}C into the medium by
cultured symbiotic *Nostoc* at intervals after separation from
the thallus (Drew and Smith, 1967*a*)

Time from separation (hours)	Percentage of total fixed ^{14}C in medium after 3 hours photosynthesis
0	46
24	41
36	12
72	15

In contrast, cells of the free-living species *Nostoc muscorum* release very little fixed carbon, ^{14}C-labelled photosynthate passing first to a soluble intracellular pool and then being converted at a constant rate into insoluble compounds, a typical pattern for free-living algae (Figure 68*b*). The fixed carbon released by the phycobiont is exclusively in the form of glucose while the major, labelled, intracellular insoluble compounds, in both symbiotic and free-living *Nostoc*, are glucose polymers. The ability of the phycobiont to release photosynthate declines rapidly with increasing time after the separation of algal cells from the thallus (Table 22). Forty-eight hours after separation, glucose is no longer released and labelled carbon is found mainly in an insoluble carbon compound, possibly a polysaccharide. Similar experiments with directly isolated symbiotic *Trebouxia* cells and cells from *Trebouxia* populations cultured axenically for considerable periods also show that release of ^{14}C-labelled photosynthate in any amount takes place only from freshly isolated cells. Here the labelled soluble compound released is not glucose but ribitol. Although a variety of soluble compounds within cells of both *Nostoc* and *Trebouxia* incorporate label during photosynthesis, only glucose from the former and ribitol from the latter are released.

The symbiotic state leads the algae to release soluble carbohydrates from their cells and once the influence of the mycobiont is removed artificially they lose this ability. In addition, incorporation of fixed carbon into simple carbohydrate decreases and instead it is incorporated into insoluble compounds (Green and Smith, 1974). Release is selective so that only a single compound passes out to be utilized by the fungus. How induction of carbohydrate release is brought about is not known, but in lichens containing *Nostoc* the mycobiont may interfere with algal cell wall metabolism. Cells of free-living *Nostoc* species, and of symbiotic *Nostoc* which have been cultured for some time, have a thick mucilaginous sheath. This sheath is lacking or much reduced in algae growing within the lichen thallus and in algae grown for short periods after separation from the thallus. Release of glucose may, therefore, be the result of a mycobiont-induced modification of mechanisms responsible for mucilage production. Hexose units that would normally be used to build up sheath material would thus be prevented from participating in polysaccharide synthesis and would pass to the extracellular environment. *Trebouxia* too has a sheath in the cultured, but not the symbiotic, form (Ahmadjian, 1959*b*). However, its intracellular insoluble compounds do not contain ribitol units, so that a different mechanism for photosynthate release must operate in lichens containing this alga.

The fate of released photosynthate in thalli has been investigated using two methods. First, in some heteromerous lichens it is possible to separate the algal layer plus some fungal material from the underlying fungal medulla. If this is done after thalli or thallus discs have been allowed to photosynthesize using labelled carbon dioxide, then extracts from the algal layer and medulla can be made, and the partition of labelled compounds between these two tissues can be determined (Smith and Drew, 1965; Drew and Smith, 1967*b*). A second method allows the identification of compounds moving from phycobiont to mycobiont without the necessity for dissection of the thallus, so that lichens which are difficult to manipulate can be studied (Drew and Smith, 1967*b*; Richardson and Smith, 1968*a*). Lichen tissues are allowed to photosynthesize in ^{14}C-labelled bicarbonate solution in the presence or absence of a high concentration of the non-radioactive form of that carbohydrate which is suspected to move from alga to fungus. Alternatively, a pulse-chase system can be used where photosynthesis takes place in ^{14}C-labelled bicarbonate solution for a time, after which the lichen material is removed to appropriate carbohydrate-rich or carbohydrate-free media for incubation. Using either system, then in the carbohydrate-rich medium, if the choice of carbohydrate is correct, movement into the fungus of the ^{14}C-labelled carbohydrate produced during photosynthesis is prevented, and it diffuses out into the medium. If the choice of carbohydrate is incorrect, or in media lacking carbohydrate, there is no such release. It is thought that release occurs because carbohydrate uptake sites on the fungus are highly specific for that particular compound which normally moves to them from the alga (Figure 69a). In the presence of high concentrations of the same, but non-radioactive, carbohydrate these uptake sites become blocked by unlabelled molecules, which are greatly in excess over labelled ones, so that the latter then appear in the medium (Figure 69b). The amount of label in the medium is taken as a measure of the amount of carbohydrate that would have moved from phycobiont to mycobiont during the equivalent period of time under normal conditions. This technique has been valuable in studying the carbohydrate metabolism of a number of autotroph—heterotroph symbioses but its degree of

(a)

1. ^{14}C-labelled carbohydrate molecules from algal photosynthesis move towards fungus.

2. Labelled molecules enter uptake sites on fungus.

3. Molecules pass into fungus and are used in fungal metabolism.

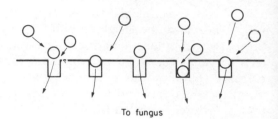

To fungus

(b)

To medium

1. ^{14}C-labelled carbohydrate molecules from algal photosynthesis move towards fungus but do so in presence of a preponderance of unlabelled molecules.

2. Excess unlabelled molecules continually occupy uptake sites on fungus.

3. Unlabelled molecules pass into fungus, the labelled molecules diffuse into the medium.

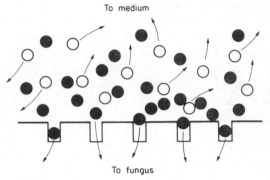

To fungus

Figure 69   Diagram of inhibition of movement from alga to fungus of a ^{14}C-labelled carbohydrate in the presence of a high concentration of unlabelled molecules of the same carbohydrate

accuracy has been questioned (Hill and Smith, 1972). An excess of external carbohydrate may distort the lichen's normal metabolic patterns so that there may be reduced net carbon fixation. Additionally, there is the possibility that competition between the two forms of carbohydrate occurs at some other point, distinct from fungal uptake sites, in the transfer pathway between the symbionts, and that carbohydrate release and uptake need not necessarily be linked processes (Drew and Smith, 1967b).

The identity of the carbohydrate released by the phycobiont has been determined for a number of lichens (Table 23). It is remarkable that in each instance only one compound appears to be moving, and that the kind of carbohydrate moving seems to depend on the type of phycobiont present in the thallus (Richardson, Hill and Smith, 1968). In lichens with blue-green algae it is glucose that moves, while in those containing green algae it is invariably a polyol. However, all blue-green algae so far studied belong to the Oscillatoriales and, since species outside this order are also found in lichens, other patterns for carbohydrate transfer may exist. Where thalli contain both a green and a blue-green alga then both glucose and a polyol are transferred to the mycobiont (Richardson, Smith and Lewis, 1967).

Lichens fall into four arbitrary categories depending on the time taken for ^{14}C-labelled compounds to be released to the medium when normal movement is prevented by inhibition (Table 23). The validity of these categories is questionable since absolute rates of movement have not been determined. Apparent rates depend on the size of the mobile carbohydrate pool, and if it is small and becomes rapidly labelled then movement will appear to be fast. It would appear to be slow if the pool were larger and consequently became labelled more slowly (Smith, Muscatine and Lewis, 1969). Specific activities of the ^{14}C-compounds involved have not been determined so that pool sizes are unknown. For this reason the apparent dependence of the rate of carbohydrate movement on the kind of alga in the thallus, rather than on the nature of the carbohydrate moving, must be viewed with caution.

Irrespective of the nature of the mobile carbohydrate it is converted, within the mycobiont, to mannitol, or occasionally to mannitol and arabitol. It may be that mannitol and, where formed, arabitol cannot be utilized by the phycobiont so that one-way movement of carbon from alga to fungus is maintained by this conversion. Polyol production seems to be a characteristic of lichenized green algae, although some free-living species synthesize polyols if subjected to droughting. The significance of polyol formation by green phycobionts is not clear but two points might be made. First, the genera of green algae so far studied are not closely related and, second, polyol production is in some way stimulated by the symbiotic association. When green phycobionts are brought into axenic culture polyol production is either reduced or ceases entirely. It has been suggested that polyols represent a form of carbohydrate which can pass through the algal cell wall and be released without involving surface enzyme systems. This, together with the fact that polyols are not, apparently, involved in insoluble polysaccharide synthesis in green algae, may have an important bearing on their role in thallus physiology (Richardson, Hill and Smith, 1968). However, there could also be an environmental basis for polyol production if this contributed towards their survival during periods of desiccation.

The presently accepted picture of carbohydrate physiology of *P. polydactyla*, and by implication that of other lichens with *Nostoc* phycobionts, may require some modification (Hill, 1972). Changes in the ^{14}C-labelling levels of various carbohydrate fractions of thallus discs have been followed using techniques of pulse feeding with subsequent incubation in a suitable buffer. The four major labelled fractions are mannitol, glucose, sugar phosphates and glucan, and the labelling of each changes with time during incubation in buffer solution (Figure 70). A balance sheet shows that a decrease in labelling of the

Table 23

Carbohydrate movement from alga to fungus in lichen species (after Richardson, Hill and Smith, 1968)

Phycobiont				
Class	Genus	Lichen species	Mobile carbohydrate	Percentage of ^{14}C moving after 3 hours during inhibition
Chlorophyceae	Trebouxia	Lecanora conizaeoides	Ribitol	
		Parmelia furfuracea	Ribitol	
		P. saxatilis	Ribitol	2–4
		Umbilicaria pustulata	Ribitol	
		Xanthoria aureola	Ribitol	
	Myrmecia	Dermatocarpon hepaticum	Ribitol	
		Lobaria amplissima	Ribitol	2–4
		L. laetevirens	Ribitol	
		L. pulmonaria	Ribitol	
	Coccomyxa	Peltigera aphthosa	Ribitol	10–15
		Solorina saccata	Ribitol	

Trentepohlia	Gyalecta cupularis	Erythritol	
	Lecanactis stenhammarii	Erythritol	
	Roccella fuciformis	Erythritol	1–2
	R. phycopsis	Erythritol	
Hyalococcus	Dermatocarpon fluviatile	Sorbitol	10–15
	D. miniatum	Sorbitol	
Cyanophyceae			
Nostoc	Collema auriculatum	Glucose	
	Leptogium sp.	Glucose	
	Lobaria scrobiculata	Glucose	
	Peltigera canina	Glucose	
	Peltigera horizontalis	Glucose	
	P. polydactyla	Glucose	20–40
	Sticta fuliginosa	Glucose	
	Sticta sp. (Cyanicaudata group)	Glucose	
	Cephalodia of Lobaria amplissima	Glucose	
	Cephalodia of Peltigera aphthosa	Glucose	
	Cephalodia of Solorina saccata	Glucose	
Calothrix	Lichina pygmaea	?Glucose/glucan	2–4
	Coccocarpia sp.	Glucose	

242

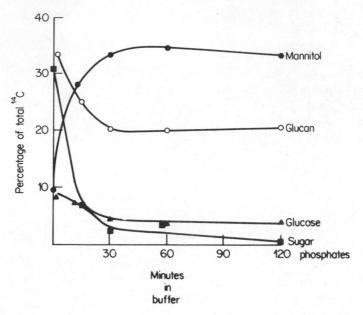

Figure 70 *Peltigera polydactyla*, redistribution of ^{14}C in various fractions after pulse feeding with labelled CO_2 followed by incubation in buffer. Modified from Hill, 1972; by permission of *New Phytologist*

last three fractions is matched by increased labelling of mannitol (Table 24). This may be interpreted as meaning that sugar phosphates, glucose, and glucan are intermediates in the carbon pathway leading from initial carbon fixation to mannitol. Most glucan is found in the algal layer where carbon transfer between the symbionts occurs. This might indicate that glucose is not released directly, but is used to form glucan outside the algal plasmalemma, and that this glucan is then hydrolysed by an extracellular fungal glucanase, the free glucose so liberated then being taken up by fungus and utilized (Figure 71). An

Table 24
Changes in levels of ^{14}C in fractions after pulse feeding with $^{14}CO_2$ (after Hill, 1972)

Fraction	Change in ^{14}C after pulse as percentage of total counts
Mannitol	+50
Glucose	−3 ⎫
Sugar phosphates	−31 ⎬ total −50
Glucan	−16 ⎭

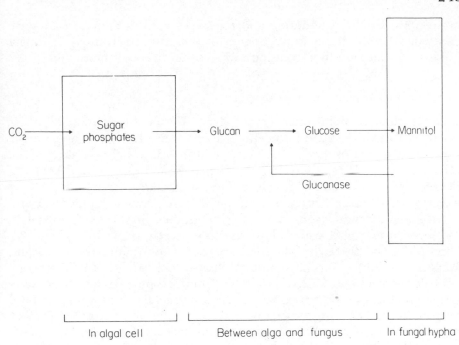

Figure 71 *Peltigera polydactyla*, possible mechanism of transfer of photo-synthate from alga to fungus. After Hill, 1972

insoluble intermediate in glucose production in thalli of *P. polydactyla* has been demonstrated using autoradiographic techniques (Peveling and Hill, 1974). This hypothesis is supported by the previously mentioned observation that freshly isolated symbiotic *Nostoc* quickly loses its ability to release glucose, and that this is correlated with an increase in the thickness of the mucilage sheath. If care is taken to remove all fragments of the mycobiont during preparation of *Nostoc* suspensions, then much less glucose is released during the initial phases of growth in axenic culture than if all fragments are not removed. This effect may be due to removal of the fungal glucanase. In the lichen *Coccocarpia*, which has *Scytonema* as the phycobiont, a glucan seems to play the same role as in *Peltigera*. Pulse feeding leads to maximum labelling of glucose occurring after maximum labelling of the glucan fraction, again implying that the latter is the first-formed compound (Hill, 1972).

Although many details are lacking, the general bases for the carbohydrate metabolism of lichens are now well known in outline. There is direct intervention by the mycobiont in the metabolism of the phycobiont and growth of the alga is severely restricted, due in part perhaps to the physical restraints which the fungus imposes. Despite this restriction, rates of photosynthesis seem to be unaffected so that conditions for the production and release of surplus carbohydrates are created. Carbohydrates move to the fungus, are transformed to fungal compounds and are then translocated to other parts of the thallus.

Fixation and Movement of Nitrogen

Lichens that contain only green algae presumably obtain all their nitrogen from the substratum, but where lichens have blue-green algae fixation of atmospheric nitrogen within the thallus takes place (Henriksson, 1951; Scott, 1956; Henriksson and Simu, 1971). In lichens with both kinds of algae fixation occurs in the internal or external cephalodia. Nitrogen fixation within external cephalodia, and nitrogen movement from these to the thallus, has been followed in *Peltigera aphthosa*, which has the blue-green alga *Nostoc* in the cephalodia and the green alga *Coccomyxa* in the thallus (Millbank and Kershaw, 1969; Kershaw and Millbank, 1970). If thalli bearing cephalodia are exposed to $^{15}N_2$, then the isotopic nitrogen content of the cephalodia at first rises but afterwards remains at a constant level (Figure 72a). The isotopic nitrogen content of thallus tissue also rises, but continues to rise, at a steady rate, indicating that nitrogen is being constantly released from cephalodial *Nostoc* cells and is passing to the thallus (Figure 72b). The rate of release of isotopic nitrogen from cephalodia after the initial build-up is presumably equal to the rate of fixation. Fixed nitrogen passing to the thallus is utilized almost exclusively by the mycobiont and only 3% of it is taken up by the *Coccomyxa* cells. Absolute rates of fixation by *Nostoc* are extremely rapid, yet there is no visible growth of the algal cells even over a period of 3 months. At 25°C the blue-green phycobiont fixes nitrogen at a rate that would permit its cells to divide every 11 hours if the alga itself assimilated the fixation products. This compares with a generation time of 19—20 hours for the free-living species *Nostoc muscorum* under optimum *in vitro* conditions. Release of fixed nitrogen may be under the metabolic control of the mycobiont through either its inhibition of algal protein synthesis, so diverting protein precursors into a secretion pathway, or by its direct action on algal membranes, making them more permeable to amino acids and oligopeptides. Nitrogen fixation in a number of lichens has been studied in relation to environmental factors (Hitch and Stewart, 1973). Thalli collected dry from the field do not fix nitrogen but will do so within an hour of being moistened, fixation rates depending on light intensity but continuing for 18—26 hours in the dark.

Although nitrogen fixation in lichens can be very rapid, in lichens with one phycobiont heterocysts may not be present in large numbers in either thalli or cephalodia (Peat, 1968; Griffiths, Greenwood and Millbank, 1972; Hitch and Millbank, 1975b). For example in *Peltigera*

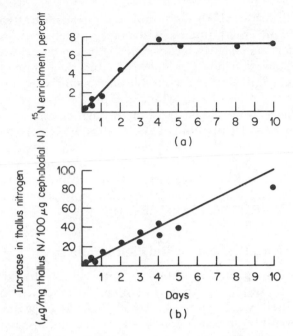

Figure 72   *Peltigera aphthosa*, fixation and movement of nitrogen. (a) ^{15}N enrichment of cephalodia exposed to 95% ^{15}N; (b) Nitrogen uptake by the thallus, measurements made after excision of cephalodia. From Millbank and Kershaw, 1969; by permission of *New Phytologist*

canina, which has a *Nostoc* phycobiont in the thallus, heterocyst frequency is 3—4% in comparison with 8—12% in the free-living alga *Anabaena cylindrica*. In contrast, in lichens with both a green and a blue-green alga, where the latter is usually located in cephalodia, heterocyst frequency is very high, being typically 20—30%. It may be that augmentation of the blue-green phycobiont's photosynthate by the green alga may stimulate heterocyst differentiation (Hitch and Millbank, 1975*a*).

Nitrogenase activity in *Nostoc* from *P. canina* has been measured and mean rates are high, being in the region of 9 nM ethylene produced per minute per milligram of algal protein, with a maximum of 14—15 nM. In the free-living species *N. muscorum* rates are 3—4 nM per minute per milligram protein, which indicates that high rates of nitrogen fixation are a characteristic feature of the lichenized condition (Millbank, 1972). For *P. canina* and similar lichens it has, therefore, been suggested that there is considerable nitrogenase activity within the vegetative cells of the phycobiont. Although the heterocyst is a specialized cell inside which favourable conditions for nitrogen fixation are maintained in environments with high oxygen tensions, free-living blue-green algae that lack heterocysts can fix nitrogen under anaerobic conditions (Stewart and Lex, 1970). In the lichen thallus the mycobiont could produce favourable nitrogen-fixing conditions by reducing the light intensity to which the phycobiont is subjected, and by maintaining a low oxygen tension through taking up oxygen from algal photosyntheis. Disruption of the thallus, which causes an increase in oxygen tension within its tissue, reduces nitrogen fixation.

It has, however, been convincingly shown that heterocyst frequency is accurately reflected in nitrogenase activity of thalli so that the action of this enzyme, and hence also nitrogen fixation, is probably restricted to the heterocysts (Hitch and Millbank, 1975b).

4. Water Relations, Respiration and Photosynthesis

Lichens have no special modifications to allow the absorption or transpiration of water so that they have little or no physiological control over their water content. Their patterns of water uptake and loss closely resemble those for gels, with uptake and loss being governed simply by those physical factors that affect evaporation (Smyth, 1934). Water is held in the spaces between hyphae in the thallus, within the fungal cell wall or, where either the phycobiont or mycobiont has a sheath, in gelatinous sheathing material. However, water uptake by living thalli is more rapid than that by dead thalli when both are incubated in humid atmospheres. This suggests that a hygroscopic mechanism is not entirely responsible for uptake, but that active uptake by living cells also plays a part, particularly in the later phases of hydration (Heatwole, 1966).

The saturated water content of thalli usually lies between 100–300% of their dry weight, this being a lower maximum water content than that found in many epiphytic cryptogamic plants (Barkman, 1958). Absorption of liquid water by most thalli when they are dry is a very rapid process, saturation being achieved in from a few seconds to about 30 minutes. This rapid influx is probably, at least initially, through the mass capillary effect of the dry cortical and medullary hyphae. Air-dry thalli can absorb atmospheric water vapour and their water content reaches a constant equilibrium value after 4–9 days. At 95% relative humidity this equilibrium value is between 30–50% of the saturated water content. This property explains the success of lichens in exploiting geographical areas where there is a periodically humid atmosphere but not enough precipitation to support growth of higher plants.

When considering respiration and photosynthesis in lichens it is difficult to separate the two phenomena because they are so closely interrelated. Nor is it easy to discuss these two processes without reference to water relations, since the degree of thallus hydration affects them profoundly. Respiration rates of thalli are generally low, being considerably lower than those rates obtaining in angiospermous leaves (Hale, 1974). As thallus water content increases so does respiratory rate and the latter follows an approximately linear course with increasing hydration. This increase continues until a maximum respiratory rate is reached, after which no further increase takes place no matter how much more the water content increases. The respiratory maximum may be achieved at any point between 40—95% of saturation depending on the species (Smyth, 1934; Harris, 1971). In heteromerous thalli medullary respiration may be lower than that of the algal layer. In general, lichen respiration rates do not show much seasonal variation and there are no clear metabolic differences between arctic and tropical lichens. The former do, however, respire slightly faster than the latter even at low temperatures, which suggests that they are to some extent cold-adapted (Scholander, Flagg, Walter and Irving, 1952).

Studies on air-dried thalli of *Peltigera polydactyla* and *Xanthoria aureola* during rehydration indicate that respiration—water uptake relationships may be complicated (Smith and Molesworth, 1973). When water is taken up by dry tissues two distinct processes occur. First, there is an immediate non-metabolic release of substantial volumes of gas comprising 75—80% carbon dioxide. This 'wetting burst' is followed by a rapid rise in respiration rate above that level found in fully hydrated thalli. This 'resaturation respiration' persists for about 2 hours in *X. aureola* and for 9 hours in *P. polydactyla* and then falls to the normal basal level. Resaturation respiration is azide- and cyanide-sensitive while basal respiration is not. In addition, resaturation respiration occurs in *P. polydactyla* only when the thallus moisture content is initially reduced below 40% of maximum. This indicates that resaturation respiration is not simply a stimulation of basal respiration, but that other pathways are brought into play when the moisture content of the thallus is reduced below a critical level.

In *P. polydactyla* total carbon dioxide loss through the wetting burst, resaturation respiration, and basal respiration during rehydration, is 3 times as high as that from wet tissues over a comparable period (Table 25). Resaturation respiration may simply be an inevitable, deleterious consequence of a lichen's ability to withstand desiccation, and might reflect a temporary loss of control over metabolism during rehydration. However, it may also serve to supply energy for the rapid reorganization of cell function or the re-establishment of those cell permeability barriers that prevent nutrient leaching during water uptake. Since respiratory activity during water uptake leads to carbon loss, then cyclic wetting and drying would lead to a drastic reduction in endogenous carbohydrate reserves. Whether this happens depends on

Table 25
Loss of CO_2 by air-dried discs of *Peltigera polydactyla* over a 9 hour rehydration period
(from Smith and Molesworth, 1973)

	CO_2 lost $\mu l/15$ discs
Wetting burst	115
Resaturation respiration	114
Basal respiration	121
Total	350
Control discs (wet)	121

the balance between respiration and photosynthesis during rehydration periods.

Photosynthetic rates vary widely between species and between individual thalli of the same species growing in different habitats. As with respiration, photosynthetic rates per unit area or unit weight of thallus are much lower than those for angiospermous leaves and several factors contribute to this (Ertl, 1951; Wilhelmsen, 1959). The chlorophyll content of lichens may be less than 25% of that of leaves and the upper fungal layers may absorb 26—43% of incident light. Lichen algae are not markedly different from other green plants in their responses to light but may make up only 3—5% of the thallus tissues.

Photosynthetic rates, like respiration rates, are related to thallus water content (Ried, 1960; Harris, 1971). However, photosynthesis often declines sharply at water contents higher than that at which the maximum photosynthetic rate is achieved. This decrease occurs at relatively low moisture levels in lichens with compact thalli or thick cortical layers, presumably because, as the cells of compacted tissue swell on hydration, they prevent efficient gaseous exchange. However, in aquatic lichens maximum photosynthetic rates are reached at 100% saturation. Optimum temperatures for photosynthesis vary, but lichens from cold zones have temperature optima below 10°C and will fix carbon dioxide even below 0°C. Tropical lichens have optima at about 20°C.

Net carbon assimilation rates obviously depend on the relative rates of respiration and photosynthesis and these may be affected differently by changes in environmental conditions, in particular changes in the availability of water. Dehydration leads to a severe reduction in photosynthesis due in part to a decrease in the light transmissivity of the lichen's upper layers. Respiration is less affected and may persist even in air-dry thalli although at a reduced rate (Ried, 1960). Fluctuating wet—dry conditions could lead to net carbon loss due to wetting burst and resaturation respiration, coupled with low photosynthetic rates during the dry period and the initial stages of rehydration. Few detailed studies on net carbon assimilation rates have been made, but that lichens are very successful organisms in a number of habitats suggests maintenance of a generally favourable carbon balance. Where such studies have been made, net carbon assimilation rates reflect seasonal variations in both photosynthetic efficiency and basal respiration rate (Harris, 1971).

All the physiological evidence indicates a delicate balance between mycobiont and phycobiont, this being an essential prerequisite for the maintenance of form and integrity of the thallus. This balance is to a large extent imposed or maintained by the kinds of environmental factors operating in characteristic lichen habitats. The major factors involved are low nutrient availability and fluctuating light—dark, wet—dry periods (Scott, 1960; Pearson, 1970). Providing thalli with nutrients, or subjecting them to continuous light, leads to thallus disintegration due to increased growth of the phycobiont. Fluctuation in availability of water seems to be essential for integrated growth of the symbionts, and it has been suggested that at low water contents only algal growth occurs, while at high water levels it is only the fungus which is able to grow. Additionally, at high water levels, most or all of the products of photosynthesis may be passed to the fungus, leading to death of the algae and breakdown of the thallus.

The established effects of interacting light—dark, wet—dry conditions in maintaining normal thallus growth and form are difficult to explain but may operate as follows (Harris and Kershaw, 1971). If the thallus is saturated and illuminated, photosynthates are passed to the fungus, the rate of fungal respiration being limited by the rate of photosynthate movement. In the dark the fungus removes stored photosynthate from the alga for respiration. At low water contents (10—15% of the maximum) and in the light, activity of the fungus is limited by lack of water, so that photosynthates are stored by the alga and growth of the latter can take place. In the dark at low water contents, stored photosynthates are respired by the alga so that the fungus remains in a substrate-limited, water-limited, physiological condition and makes little growth. In this way it is possible for the alga to grow at lower thallus water contents than the fungus.

5. Other Characteristics

Lichen thalli grow very slowly and their development is influenced by environmental conditions to a greater degree than that of most terrestrial fungi. Because of their slow growth, and consequent lack of long-term observations, little or nothing is generally known about growth dynamics of thalli, their development rate in relation to age, and the contribution of different parts of the thallus to total growth (Ahmadjian, 1967). Reduced growth might obviously be due to slow intrinsic growth rates of the symbionts, but it could also be imposed by the nutrient-poor nature of most lichen substrates. The nature of the physical organization of thalli also usually means that they are rapidly affected by environmental changes, particularly by the amount of water available, and periods of growth have been found to coincide with periods of rainfall (Armstrong, 1973). Factors such as this allow lichens only relatively brief, discontinuous, periods during which their metabolic processes can function at an optimum level.

Although lichens are very sensitive to atmospheric pollutants, particularly sulphur dioxide which causes chlorophyll destruction, they are extremely resistant to environmental extremes especially when in a state of dehydration (Kappen and Lange, 1970; Puckett, Niebeer, Flora and Richardson, 1973). For example, cooling to below $-70°C$ does not impair subsequent activity of either the mycobiont or phycobiont. Even tropical lichens have a high degree of cold resistance so that this may be an innate property of the lichen association. Lichens are also heat resistant to temperatures in excess of $70°C$ and in nature in exposed situations thallus temperatures may be in excess of $50°C$. Slow growth rates together with high resistance to water and heat stress confer a durability on lichens that is reflected in the great age of some thalli which may be up to 4,500 years old. However, many lichens have a maximum thallus size and a finite life span of 10—20 years (Hale, 1974).

Another characteristic of lichens is the production by the mycobiont of organic compounds which form extracellular deposits on the hyphal surfaces. These compounds are nearly all unique to lichen fungi, are not water-soluble, and are not formed when the fungi are grown in axenic culture. A great variety of aliphatic and aromatic compounds, mainly weak phenolic acids, have been found in lichens and many are pigments (Ahmadjian, 1967; Shibata, 1958, 1963; Culberson, 1969). It is difficult to assign a metabolic role to these compounds and they seem not to be reabsorbed by the hyphae after deposition. They may, however, still have an importance in the physiology of the thallus. Lichen acids and pigments reduce the amount of light reaching the algal layer and presumably filter out harmful wavelengths. Crystalline deposits also make thalli unwettable, prevent excessive water uptake and help to regulate internal water levels. Atranorin, a derivative of phenolcarboxylic acid, may act as a fluorphor. It fluoresces at 425 nm which is coincident with an absorption peak for chlorophyll in the blue part of the spectrum. It may, therefore, increase the ability of the phycobiont to utilize light of the shorter wavelengths (Rao and Leblanc, 1966). Many lichen compounds have antibiotic properties both against bacteria and other fungi, but whether this characteristic is important in nature is not known. Lichens can absorb and accumulate considerable amounts of minerals from their substrates, particularly trace elements such as zinc. The production of lichen acids may affect the mobilization of these when lichens are growing on rocky substrates or bring about their chelation within the thallus.

6. Origins and Evolution

The origin and evolution of lichens has been the subject of a great deal of speculation (Ahmadjian, 1960; Scott, 1964; Smith, Muscatine and Lewis, 1969). In the well-developed thallus, the mycobiont is biotrophic and both symbionts occupy ecological niches in which presumably neither partner could grow successfully if alone. Most algae release extracellular metabolites during normal growth so that saprotrophic, free-living fungi can develop loose ecological associations with them, such associations, apparently, being common. Present-day lichens probably originated in this manner and from this simple, basic situation there are several possible ways in which a more intimate physical and physiological association could have evolved. One possibility is that the first step was the development of a state of balanced necrotrophy between the fungus and the alga. Algal cells, particularly senescent ones, would have been invaded and killed, but their loss would have been compensated for by algal cell division. Such a condition may presently be reflected in those poorly organized lichens where frequent penetration of the phycobiont occurs.

Biotrophy could have evolved from balanced necrotrophy if the alga developed tolerance to fungal invasion. It could equally, and simultaneously, have developed directly from the original, loose saprotrophic association. In the latter case extracellular metabolites, in particular perhaps polysaccharide sheathing materials, would have been utilized, at first possibly relatively slowly. Later the fungus might have become more and more dependent on these materials, so inducing their more rapid synthesis by the alga. In this way metabolic control of their production would have come more and more under the control of the fungus. Concomitant with this would have been the development by the fungus of a means of restricting algal growth, and of a more complex control of algal metabolism, for example the loss of fixed nitrogen from blue-green algal cells.

In those associations where polyols move from alga to fungus, but are not involved in algal cell wall or sheath synthesis, the fungi may have become ecologically dependent on this form of carbon for growth. Some lichen fungi may have become reliant on other algal compounds, not necessarily major metabolites, for either growth or spore germination (Scott, 1964).

252

There is little that is certain among all the speculation except, perhaps, that lichen associations did not arise from fungi seeking out suitable algal partners that might have enhanced their ecological success, but that a great number of indiscriminate partnerships developed and are continuing to do so. Only those associations containing algae with physiological properties that allow them to be exploited without their rapid death will be successful.

Bibliography

Ahmadjian, V., 1959a. *Mycologia,* **51**: 56—60.
Ahmadjian, V., 1959b. *Svensk bot. Tidskr.,* **63**: 250—254.
Ahmadjian, V., 1960. *Bryologist,* **63**: 250—254.
Ahmadjian, V., 1961. *Bryologist,* **64**: 168—179.
Ahmadjian, V., 1962. *Am. J. Bot.,* **49**: 277—283.
Ahmadjian, V., 1964. *Bryologist,* **67**: 87—98.
Ahmadjian, V., 1966. *Science,* **151**: 199—201.
Ahmadjian, V., 1967. *The Lichen Symbiosis,* Blaisdell, Toronto and London.
Ahmadjian, V., Heijkilä, H., 1970. *Lichenologist,* **4**: 259—267.
Ahmadjian, V. and Jacobs, J. B., 1970. *Lichenologist,* **4**: 268—270.
Armstrong, R. A., 1973. *New Phytol.,* **72**: 1023—1030.
Barkman, J. J., 1958. *Phytosociology and Ecology of Cryptogamic Epiphytes,* Van Gorcum, The Hague.
Bednar, T. W. and Holm-Hansen, O., 1964. *Pl. Cell. Physiol.,* **5**: 297—303.
Bednar, T. W. and Smith, D. C., 1966. *New Phytol.,* **65**: 211—220.
Ben-Shaul, Y., Paran, N. and Galun, M., 1969. *J. Microscopie,* **8**: 415—420.
Chervin, R. E., Baker, G. E. and Hohl, H. R., 1968. *Can J. Bot.,* **46**: 241—245.
Culberson, C. F., 1969. *Chemical and Botanical Guide to Lichen Products,* University of North Carolina Press.
Drew, E. A. and Smith, D. C., 1967a. *New Phytol.,* **66**: 379—388.
Drew, E. A. and Smith, D. C., 1967b. *New Phytol.,* **66**: 389—400.
Durrell, L. W., 1967. *Mycopath. Mycol. appl.,* **31**: 273—280.
Ertl, L., 1951. *Planta,* **39**: 245—270.
Fox, C. H., 1966. *Physiologia Pl.,* **19**: 830—839.
Fox, C. H., 1967. *Physiologia Pl.,* **20**: 251—262.
Galun, M., Paran, N. and Ben-Shaul, Y., 1970a. *New Phytol.,* **69**: 599—603.
Galun, M., Paran, N. and Ben-Shaul, Y., 1970b. *Protoplasma,* **69**: 85—96.
Green, T. G. A. and Smith, D. C., 1974. *New Phytol.,* **73**: 753—766.
Griffiths, B. H., Greenwood, A. D. and Millbank, J. W., 1972. *New Phytol.,* **71**: 11—13.
Gross, M. and Ahmadjian, V., 1966. *Svensk bot. Tidskr.,* **60**: 74—80.
Hale, M. E., 1958. *Bull. Torrey bot. Club.,* **8**: 182—187.
Hale, M. E., 1974. *The Biology of Lichens,* Arnold, London.
Harris, G. P., 1971. *J. Ecol.,* **59**: 441—452.
Harris, G. P. and Kershaw, K. A., 1971. *Can. J. Bot.,* **49**: 1367—1372.
Heatwole, H., 1966. *Mycologia,* **58**: 148—156.
Henriksson, E., 1951. *Physiologia Pl.,* **4**: 542—551.
Henriksson, E., 1961. *Physiologia Pl.,* **14**: 813—817.
Henriksson, E., 1964. *Svensk bot. Tidskr.,* **58**: 361—370.
Henriksson, E. and Simu, B. 1971. *Oikos,* **22**: 119—121.
Hill, D. J., 1972. *New Phytol.,* **71**: 31—39.
Hill, D. J. and Smith, D. C., 1972. *New Phytol.,* **71**: 15—30.

Hitch, C. J. B. and Millbank, J. W., 1975a. *New Phytol.*, 75: 239—244.
Hitch, C. J. B. and Millbank, J. W., 1975b. *New Phytol.*, 74: 473—476.
Hitch, C. J. B. and Stewart, W. D., 1973. *New Phytol.*, 72: 509—524.
Jacobs, J. B. and Ahmadjian, V., 1973. *New Phytol.*, 72: 155—160.
Kappen, L. and Lange, O. L., 1970. *Lichenologist*, 4: 289—293.
Karling, J. S., 1934. *Ann. Bot.*, 48: 823—855.
Kershaw, K. A. and Millbank, J. W., 1970. *New Phytol.*, 69: 75—79.
Kohlmeyer, J., 1968. *Phytopath. Z.*, 63: 341—363.
Kohlmeyer, J. and Kohlmeyer, E., 1972. *Botanica mar.*, 15: 109—112.
Kremer, B. P., 1973. *Marine Biol.*, 22: 31—35.
Millbank, J. W., 1972. *New Phytol.*, 71: 1—10.
Millbank, J. W. and Kershaw, K. A., 1969. *New Phytol.*, 68: 721—729.
Moore, R. T. and McAlear, J. H., 1960. *Mycologia*, 52: 805—807.
Paran, N., Ben-Shaul, Y. and Galun, M., 1971. *Arch. Mikrobiol.*, 76: 103—113.
Pearson, L. G., 1970. *Am. J. Bot.*, 57: 659—664.
Peat, A., 1968. *Arch. Mikrobiol.*, 61: 212—222.
Peveling, E., 1973. *New Phytol.*, 72: 343—345.
Peveling, E. and Hill, D. J., 1974. *New Phytol.*, 73: 767—769.
Plessl, A., 1963. *Öst. bot. Z.*, 110: 194—269.
Puckett, K. J., Niebeer, E., Flora, W. P. and Richardson, D. H. S., 1973. *New Phytol.*, 72: 141—154.
Rao, D. N. and LeBlanc, F., 1966. *Bryologist*, 69: 69—75.
Richardson, D. H. S., Hill, D. J. and Smith, D. C., 1968. *New Phytol.*, 67: 469—485.
Richardson, D. H. S. and Smith, D. C., 1968a. *New Phytol.*, 67: 61—67.
Richardson, D. H. S. and Smith, D. C., 1968b. *New Phytol.*, 67: 69—77.
Richardson, D. H. S., Smith, D. C. and Lewis, D. H., 1967. *Nature, Lond.*, 214: 879—882.
Ried, A., 1960. *Biol. Zbl.*, 79: 129—151.
Roskin, P., 1970. *Arch. Mikrobiol.*, 70: 176—182.
Scholander, P. F., Flagg, W., Walters, V. and Irving, L., 1952. *Am. J. Bot.*, 39: 707—713.
Scott, G. D., 1956. *New Phytol.*, 35: 111—116.
Scott, G. D., 1960. *New Phytol.*, 59: 374—381.
Scott, G. D., 1964. *Z. allg. Mikrobiol.*, 4: 326—336.
Shibata, S., 1958. In W. Ruhland (ed.), *Handbuch der Pflanzenphysiologie*, p. 560—623, Springer, Berlin
Shibata, S., 1963. In K. Paech and M. V. Tracey (eds.), *Modern Methods of Plant Analysis*, Vol. 6, p. 155—193, Springer, Berlin.
Smith, D. C. and Drew, E. A., 1965. *New Phytol.*, 64: 195—200.
Smith, D. C. and Molesworth, S., 1973. *New Phytol.*, 72: 523—533.
Smith, D., Muscatine, L. and Lewis, D., 1969. *Biol. Rev.*, 44: 17—90.
Smyth, E. S., 1934. *Ann. Bot.*, 48: 781—818.
Stewart, W. D. and Lex, M., 1970. *Arch. Mikrobiol.*, 73: 250—260.
Sutherland, G. K., 1915. *New Phytol.*, 14: 33—42.
Walker, A. T., 1968. *Am. J. Bot.*, 55: 641—649.
Webber, F. C., 1967. *Trans. Br. mycol. Soc.*, 50: 583—601.
Webber, M. M. and Webber, P. J., 1970. *Can. J. Bot.*, 48: 1521—1524.
Wetmore, C. M., 1973. *New Phytol.*, 72: 535—538.
Wilhelmsen, J. B., 1959. *Bot. Tidsskr.*, 55: 30—36.

Interfungal Symbioses

Chapter 14

Antagonism between Fungi

As well as having antagonistic, neutral and mutualistic associations with plants and animals, many fungi have these kinds of relationships with other fungi (De Vay, 1956; Barnett, 1963, 1964; Boosalis, 1964; Barnett and Binder, 1973). Neutral and mutualistic associations between fungi can be simulated in the laboratory but it is not known to what extent these obtain in natural ecological situations. In contrast, many antagonistic associations are well known from nature and the physiological bases for some of them are at least partly understood. Examples of such antagonistic associations are found in all the major groups of fungi, including the Myxomycota and lichen fungi, and it is probable that those so far documented represent only a small fraction of the total that exist (Buller, 1924; Ainsworth and Bisby, 1971).

A wide spectrum of fungi is involved in antagonistic interfungal symbioses, ranging from facultative, unspecialized, or 'casual' species to ecologically obligate and highly specialized fungi. Two terms, 'myco-parasitism' and 'hyperparasitism' have been used to describe the state of antagonistic symbiosis between fungi. Both are unsatisfactory, the second particularly so as it should only be applied, if at all, strictly to those situations where a fungus is antagonistic towards another which is itself antagonistic towards a third organism. 'Mycoparasitism' is widely employed, although it has also been used to refer to the antagonistic effects of fungi on higher plants, and it is arguable on etymological grounds that all diseases caused by fungi might be said to be caused by 'mycoparasites'. 'Mycoparasitism' is, in addition, unsatisfactory because it is frequently applied to associations where antagonism in the broadest possible ecological sense is involved, that is during simple competition for available substrates. Furthermore, some of the more specialized manifestations of 'mycoparasitism' have pronounced mutualistic characteristics. Although in the past it may have provided a useful blanket-term there is a strong case for not perpetuating its usage, at least with respect to interfungal symbioses.

Fungi that are antagonistic towards other fungi may be divided into two groups depending on whether they are necrotrophic or biotrophic. There are, however, examples of fungi where both kinds of nutritional behaviour occur during different phases of their association with their hosts, while in other instances there is insufficient information to allow a fungus to be placed firmly in either group.

1. Necrotrophic Fungi

The great majority of known nectrotrophic species are facultative symbionts, being frequently common soil fungi, for example *Rhizoctonia solani, Trichoderma viride, Gliocladium* and *Verticillium* species, with well-developed saprotrophic abilities (Butler, 1957; Barnett and Lilly, 1962; Dayal and Barron, 1970; Dennis and Webster, 1971). They are easily grown in axenic culture, have relatively rapid growth rates, no special nutrient requirements, and wide host ranges. Some species are antagonistic to fungi that are themselves antagonistic towards other fungi.

When growing in dual culture within the zone of influence of a host fungus, the lateral branches of the antagonist show directed growth towards host hyphae. These laterals contact host hyphae and may coil around them, sometimes in contact sometimes not (Figure 73a, b). Hyphae of the antagonist may then penetrate the host's cell wall and

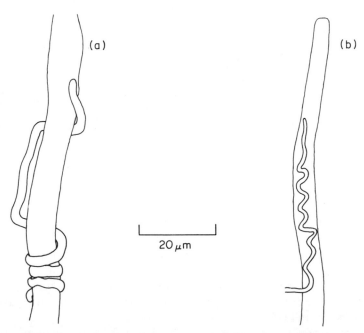

Figure 73   Necrotrophic fungi: (a) hypha of *Trichoderma polysporum* coiled round a hypha of *Rhizoctonia solani*; (b) hypha of *Verticillium* within a hypha of *R. solani*. (a) From photographs by Dr. C. Dennis

grow rapidly within the cytoplasm. Penetration of the host is not, however, a characteristic of all necrotrophic antagonists. Death of host hyphae, or some of their constituent cells, occurs but not all contacted hyphae are necessarily killed, although their growth may be severely impaired. In *Verticillium psalliotae* on *Rhopalomyces*, growth of the antagonist is entirely within host hyphae and conidia, morphogenesis of the latter being apparently unaffected, so that the antagonist is disseminated inside viable host spores.

The biochemical basis for tropic growth of lateral hyphae is not understood, but they may be responding to a nutrient or metabolite gradient arising from the host. It is also not clear how such coiling is directed. Coiling does not take place in all host—antagonist combinations and sometimes occurs around fine glass threads, so that it may be a partly thigmotropic, partly chemotropic response to the presence of host hyphae (Butler, 1957; Dennis and Webster, 1971). Factors bringing about the death of host cells are quite local in their action and do not appear to be highly diffusible, which implies that they are not antibiotics or toxins but that they are possibly enzymes. However, in associations involving some isolates of *Trichoderma viride*, vacuolation and coagulation factors act upon host cells over distances of 3—5 mm. Not all host hyphae may be equally susceptible to the effects of antagonists and age sometimes confers resistance to penetration. Penetrating hyphae may also be walled-off by activity of host cytoplasm in the immediate area of penetration. Sometimes, after gaining entry to host cells, penetration hyphae become severely lysed, reflecting a reversal of the normal direction of the antagonistic relationship.

It is an open question as to whether the possession of the kinds of antagonistic mechanisms described here confers an advantage on facultative necrotrophs when they are growing naturally in soil or organic substrates. Other properties, for example rapid mycelial growth rates and rapid germination of spores, enable them to efficiently exploit available nutrients in organic matter in the face of competition from other saprotrophic fungi. Further important advantages would accrue from direct antagonism only if there were significant nutrient transfer from host to antagonist, or if the saprotrophic activities of the host were drastically reduced. There is little positive evidence that either occurs. Nutrients may obviously be absorbed by penetrating hyphae but there is only limited evidence that leakage of nutrients from host cells is induced in the absence of penetration (Dennis and Webster, 1971). Even if leakage does take place, its contribution to the entire nutrition of a vigorously growing antagonist with good saprotrophic ability is likely to be small. In only one species, *Pythium acanthicum*, does there seem to be any kind of dependence upon the host (Haskins, 1963). This fungus requires fat-solvent compounds, probably sterols, for sexual reproduction. These are not present in its own mycelium but are present in that of many of its hosts. However, in most other instances there is nothing to suggest that an antagonist grows any better in the presence of a host fungus than it does in its absence. In addition, although in restricted, artificial systems, death of all host hyphae can be brought about by antagonistic fungi, in nature the bulk of the host's somatic material will be preserved if host hyphae are capable of a growth rate approaching or exceeding that of the anatagonist. This situation contrasts strongly with that involving saprotrophic fungicolous fungi which inhabit ecological niches where senescent fungal fruit body tissues provide the main substrate, there being here no possibility of the host evading exploitation.

It should be pointed out that fungi removed from ecological niches where antagonistic activity of the type so far discussed might be expected to be unnecessary, are frequently strongly antagonistic towards a wide range of other fungi in laboratory culture (Griffith and Barnett, 1967). The lignicolous Basidiomycotina *Polyporus adustus, Pleurotus ostereatus, Coriolus versicolor* and *Schizophyllum commune* are species that exhibit this kind of behaviour, yet they occupy a substrate from which most other fungi are excluded by virtue of its resistance to enzymic breakdown. The antagonistic habit in such a situation would, therefore, seem to be superfluous except in special circumstances, although a high carbon–nitrogen ratio in the substrate (a condition obtaining in dead wood) seems to increase the antagonistic propensities of *P. adustus*. The implication is that this species may, under certain conditions, supplement its nitrogen supply by exploiting a host fungus.

Only two nectrotrophic species, both Hyphomycetes in the Sphaeropsidales, have a markedly restricted host range and a concomitant restriction in their habitat range. *Darluca filum* is geographically cosmopolitan but is found only in rust pustules, particularly urediosori and teliosori. Pustule mycelium and spores are destroyed although there is no penetration of host hyphae (Barnett and Binder, 1973). It is easily cultured and grows on chemically-defined media where it utilizes a wide range of carbon and nitrogen compounds, including inorganic nitrogen sources, and shows a partial requirement for biotin (Nicolas and Villanueva, 1965; Bean, 1968). It appears to have no special nutritional requirements that might account for its natural restriction to rust pustules. Many species of the Sphaeropsidales are leaf-inhabiting fungi, and it is possible that *D. filum* has evolved from a leaf-inhabiting ancestor but, having lost those attributes necessary for the colonization of leaf tissue, it has become dependent on rust fungi which do possess those attributes. The host fungus may affect *D.filum* in ways other than simply providing nutrients, since compounds from urediospores have been found to increase conidium germination, mycelial growth, and pycnidium formation (Rambo and Bean, 1970; Swendsrud and Calpouzos, 1970).

The second hyphomycetous species, *Ampelomyces quisqualis*, is found in nature associated with the mycelium of powdery mildews. However, it can be induced to attack species in the Mucorales some of which may themselves be antagonistic towards other fungi (Linnemann, 1968). It grows well in axenic culture and, although its nutrient requirements have not been determined, it is possible that it is restricted in nature to powdery mildew mycelium for reasons similar to those determining the restriction of *D.filum*. *A.quisqualis* can form pseudoparenchymatous pads between the cuticle and epidermis of the leaf when the latter becomes moribund due to the activity of the powdery mildew. Penetration of epidermal cells and intercellular growth have both been observed, so that it may have some ability to exploit the still-living leaf. It can certainly survive as a saprotroph in dead leaf tissue (Emmons, 1930).

2. Biotrophic Fungi

In contrast to the necrotrophs, biotrophic fungi are typically highly specialized antagonists, and probably do not exist in nature in the vegetative state unless associated with a suitable host. They also tend to belong to discrete, and sometimes small, taxonomic groups and probably have relatively narrow natural host ranges, although these may widen in laboratory conditions. Three series of biotrophic fungi are known, each having distinct taxonomic and behavioural characteristics. The three groups are the symbiotic Chytridiomycetes, the filamentous contact symbionts, and the filamentous haustorial symbionts (Barnett and Binder, 1973).

Symbiotic Chytridiomycetes

All symbiotic chytrids so far described seem to be biotrophic although during the final stages of growth they may become necrotrophs. They are usually restricted to other chytrids, which may themselves be attacking other fungi, or to aquatic and semi-aquatic Oomycetes, particularly members of the Saprolegniales (Karling, 1942, Willoughby, 1956; Karling, 1960; Held, 1973a,b). Development of the symbiont may take place either entirely within the host's cytoplasm or outside it on the thallus surface, rhizoids or hyphae. In the latter situation the host's cytoplasm is exploited by the biotroph by means of rhizoidal filaments that may sometimes be modified to form sac-like haustoria. Effects on the host may vary from its more or less rapid death to non-fatal changes, such as localized cell wall growth to form swellings, or to increased septation of host hyphae. Few experimental investigations have been reported and it is possible that details of antagonist—host interactions are more complex than their morphological simplicity would suggest. For example, mycelia of *Saprolegnia* and *Isoachlya* are susceptible to *Olpidiopsis incrassata* only before asexual or sexual reproduction of the host is initiated (Slifkin, 1961). It seems that resistance is determined by the host's hormonal status, an apparently unique situation in inter-fungal symbioses.

Filamentous Contact Biotrophs

Relatively few species of contact biotrophs have been described or studied in detail (Table 26). All are Hyphomycetes and three of them, *Calcarisporium parasiticum, Gonatobotrys simplex* and *Gonatobotryum fuscum*, are notable for their narrow host range. *Calcarisporium parasiticum* is confined to fungi with pycnidial imperfect stages, *Gonatobotrys simplex* to the hyphomycetous genera *Alternaria* and *Cladosporium,* and *Gonatobotryum fuscum* to *Ceratocystis* species or to genera with a similar asexual form to *Ceratocystis.* Two further species not in Table 26 are probably contact biotrophs although direct evidence is lacking. These are *Calcarisporium arbuscula*, which is found in basidiocarps of *Russula* and *Lactarius* species, and the similarly

Table 26

Contact biotrophs with their natural hosts and other host fungi determined under laboratory conditions

	Natural host	Host fungi under laboratory conditions
Calcarisporium parasiticum (Barnett, 1958)	*Dothiorella quercina* (Sphaeropsidales)	*Physalospora* and related fungi (Sphaeriales)
Gonatobotrys simplex (Whaley and Barnett, 1963)	*Alternaria* species	*Cladosporium*
Gonatobotryum fuscum (Shigo, 1960)	*Ceratocystis* species	*Graphium* and *Leptographium*
Gonatorrhodiella highlei (Gain and Barnett, 1970)	*Nectria coccinea* var. *faginata*	*Cladosporium* and *Tritirachium*
Stephanoma phaeospora (Butler and McCain, 1968; Rakuidhyasastra and Butler, 1973).	*Fusarium*	Ascomycotina and Hyphomycetes, *Ustilago*

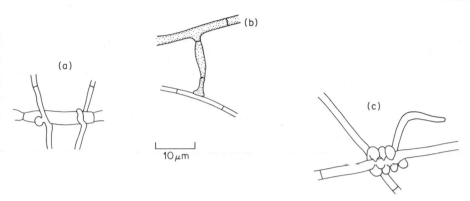

Figure 74   Contact biotrophs: (a) contact of host hypha by two hyphae of *Gonatobotrys simplex*; (b) contact by *Gonatobotryum fuscum* of its host *Graphium* (bottom); (c) indentation of host hypha by branches of *Stephanoma phaeospora*, (a) From Whaley and Barnett, 1963, by permission of *Mycologia*; (b) from a photograph in Shigo, 1960, by permission of *Mycologia*; (c) from Rakuidhyasastra and Butler, 1976, by permission of *Mycologia*.

fungicolous species *Calarisporium pallidens* (Tubaki, 1955; Watson, 1965a).

In this group of biotrophs host hyphae are contacted, but not penetrated, by specialized, short, lateral hyphae. They may either simply touch the surface of the host hypha, or become flattened upon it, or grow around it (Figure 74a-c). In *Stephanoma phaeospora* a wide range of hyphal morphology occurs, the most complicated arrangement being a series of short, contiguous branches which clamp around the host hypha causing invagination and localized thickening of the host's cell wall (Figure 74c). It is possible that in some species growth of laterals towards host hyphae does not take place at random, but is in some way directed, and that the type of contact structure formed may be produced in response to a thigmotropic stimulus. Growth of the host's lateral hyphae may also be directed by the biotroph, resulting in end to end contact between the hyphae of biotroph and host. In *C. parasiticum* and *G. simplex* tropic growth of host hyphae is induced when spores of these two species germinate close to them. This host response may be related to, and could compensate for, the normally slow germ tube growth in these biotrophs. However, in other species, where germ tube elongation is similarly slow, no such host response occurs (Barnett and Binder, 1973).

When contact is made with the host there is no obvious effect on the cytoplasm within host hyphae. At the gross level the only effect of biotroph—host contact that has so far been established is a reduction in the host's growth rate in dual culture (Barnett, 1964). Whether this is due to competition between the two organisms for nutrients or to more direct effects remains undetermined. Dual cultures of host and biotroph can be easily maintained, but it is unlikely that simple nutritional studies on dual systems can suffice to gain insight into the bases for contact biotrophy, since there are obvious difficulties involved in the interpretation of results. Where meaningful information is available this has been obtained mainly from attempts to bring contact biotrophs into axenic culture (Whaley and Barnett, 1963; Gain and Barnett, 1970; Calderone and Barnett, 1972; Rakuidhyasastra and Butler, 1973).

All the species listed in Table 26 can be grown in axenic culture but only on media containing mycotrophein, a water-soluble factor or factors extractable form host mycelia. *Gonatorrhodiella highlei* also requires other factors present within, but not extractable from, host hyphae (Gain and Barnett, 1970). Curiously, conidia of *Gonatobotryum fuscum* and *Stephanoma phaeospora* do not require mycotrophein for germination and will do so freely in distilled water. Mycotrophein has not been characterized but it is of low molecular weight, thermostable, active at levels as low as 1 ppm, and is not identifiable with any major nutrient or B-group vitamin. Nutrient requirements of biotrophs growing on media containing mycotrophein vary with the species, there being no common pattern. For example *C. parasicum, G. fuscum* and *G. highlei* will utilize single amino acids as sole nitrogen sources, while *Gonatobotrys simplex* has complex nitrogen requirements. In addition *G. fuscum* requires abnormally high levels of biotin and thiamine.

Ascomycotina, Hyphomycetes and Basidiomycotina, but not Mastigomycotina or Zygomycotina, contain species that can produce mycotrophein or mycotrophein-like growth factors. Detailed studies on *S. phaeospora* show that, although all its hosts produce mycotrophein, this factor is also produced by non-host species, and extracts from the latter support good growth of the biotroph (Rakuidhyasastra and Butler, 1973). Some host and non-host fungi can secrete mycotrophein into the medium, so that enhanced saprotrophic growth of the antagonist occurs in close proximity to these species without the necessity for contact. It might be expected that possesion of mycotrophein-synthesizing systems determines whether a particular fungus will act as a host or not, but this is clearly not the case, and what kinds of factors do determine this remain unknown.

Nutritional and other investigations on contact biotrophs have not indicated clearly why any particular species has a requirement for association with a host fungus, nor what compounds are obtained by the antagonist either from the medium or from the host. Contact hyphae are considered to act as bridges for passage of nutrients from host to biotroph and to provide areas on host hyphae where increased permeability of host membranes might be induced. There is no positive evidence that contact hyphae function in either way, and *S. phaeospora* at least does not require contact to make good growth. Proximity of a host, provided that mycotrophein is secreted into the environment by it, is sufficient. In terms of major nutrients, how much the host contributes to growth and how much is obtained by the biotroph from the substrate is not known. It is likely that there is no general pattern for all species but that a range of different host—biotroph interactions exists. At the simplest, as in *S. phaeospora*, the host provides mycotrophein through either secretion or contact, and all major nutrients are probably obtained from the environment. For other species the situation may be more complex, with dependence on the host for both mycotrophein and major nutrients.

The interaction between a biotroph and its host is not invariably antagonistic and under some conditions that between *G. fuscum* and a species of *Graphium* is mutualistic (Barnett, 1968). *G. fuscum* is heterotrophic for biotin but can synthesize pyridoxine, while *Graphium* can synthesize biotin but requires pyridoxine. In dual culture in media deficient in both biotin and pyridoxine, growth of the combined mycelia is at first slow but gradually accelerates as each species supplements the vitamin requirements of the other. In ageing dual cultures, *Graphium* hyphae may attempt to penetrate those of *G. fuscum*, although it is usually excluded by the formation of resistance sheaths around the invading hyphae. Another mutualistic association is possibly that between *Calcarisporium arbuscula* and its hosts. This species is in nature restricted to fruit bodies of *Russula* and *Lactarius*, possibly because of its low competitive saprotrophic ability (Watson, 1955, 1965a). It differs from other contact biotrophs in that it can be grown in axenic culture in media lacking host extracts. When growing within basidiocarps it does no obvious damage and produces an antibiotic, calcarin, which confers upon the host's fruit body resistance to attack by other fungicolous fungi. The biotroph prolongs the existence of the host and at the same time preserves its own abode from invasion by competing fungi.

Filamentous Haustorial Biotrophs

Haustorial biotrophy has so far only been found in fungi belonging to merosporangial families of the Mucorales, although it is entirely possible that similar species may exist outside these groups. Their host ranges are, in general, restricted to other members of the Mucorales but some may attack Ascomycotina and Hyphomycetes. One species, *Dispira simplex*, is mainly restricted to *Chaetomium* species but isolates have been found on *Mucor* and *Circinella*. (Dobbs and English, 1954; Indoh, 1965; Brunk and Barnett, 1966).

Spores of haustorial biotrophs germinate in response to one or more factors diffusing from host hyphae, and their germ tubes show tropic growth towards the host. Factors stimulating germination are also produced by non-host fungi and are present in a wide variety of plant

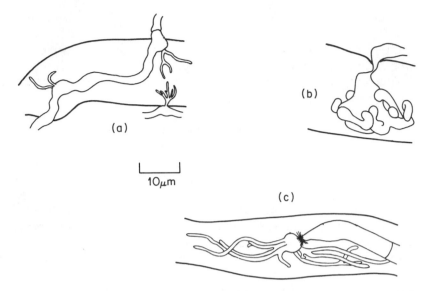

Figure 75   Haustoria of biotrophs: (a)   *Piptocephalis lepidula*;   (b)
Dimargaris cristalligena;   (c) *Dispira cornuta*. From Benjamin, 1959; by
permission of J. Cramer

materials (Berry and Barnett, 1957). Similar tropic responses of mature
vegetative hyphae may occur. On contacting the host the germ tube or
hyphal tip usually swells to form a distinct appressorium and from this
the host's cell wall is breached by a fine penetration tube. On entering
the host the tip of this swells, grows, and branches to form a small
haustorial organ. The ultimate haustorial branches may resemble
vegetative hyphae or may differ from them, being either swollen, or
fine and thread-like (Figure 75). Penetration and haustorium formation
may cause the penetrated host cell to greatly enlarge and can also
induce intense lateral branching to produce a growth resembling a
witches broom, (Brunk and Barnett, 1966; England, 1969; Hunter and
Butler, 1975).

Changes occur in the host's cytoplasm and organelles close to the
haustorium (Armentrout and Wilson, 1969; Manocha and Lee, 1971).
Tubular endoplasmic reticulum appears and, although nuclei are
unaffected, host mitochondria swell, become spherical, or fragment.
The significance of these changes is not clear. Electron-microscope
studies on *Piptocephalis virginiana* have shown that its haustoria are
similar in structure to those formed by biotrophic fungi that exploit
higher plants, except that no papilla is formed around the neck of the
haustorium by the host cell wall. Experiments using ^{3}H-labelled
N-acetylglucosamine have shown that the electron-dense matrix around
the haustorial wall, also typical of most fungal haustoria so far studied,

is probably a host structure and is not formed by *P. virginiana* itself (Manocha, 1972).

It has frequently been noted that, when growing in dual culture with their hosts, a high carbon to nitrogen ratio in the medium is unfavourable to development of haustorial biotrophs and that artificial maintenance of high carbon levels in media prevents growth (Ayers, 1935; Berry, 1959; Shigo, Anderson and Barnett, 1961; Brunk and Barnett, 1966; Kurtzman, 1968). The degree of infection of *Mortierella pusilla* by *P. virginiana* has been found to be determined by the amount of soluble nitrogen in the host's mycelium, a high nitrogen level being unfavourable to the biotroph. The effects upon the biotroph of different carbon—nitrogen ratios in the medium, or of the host's endogenous nitrogen content, must be complex, but in some instances it may be simply the mechanical resistance of the host that is being directly influenced (England, 1969). For example, *P. virginiana* attacks young but not old hyphae of *Phycomyces blakesleeanus* but will penetrate those of *Choanephora cucurbitarum* whatever their age. Tests of cell wall strength indicate that those of *C. cucurbitrarum* are much weaker than those of *P. blakesleeanus*, although the two fungi do not differ much with respect to cell wall thickness or chitin content, and that old hyphae of *P. blakesleeanus* are many times stronger than young ones. Provision of a medium rich in nitrogen might delay cell wall maturation and so directly influence susceptibility, as well as providing a high endogenous nitrogen level in the host. There is also evidence that the nature of the nitrogen source provided in the medium has important effects on biotrophs growing in dual culture with their hosts (Berry, 1959). *P. virginiana* will infect *M. pusilla* when the sole nitrogen source is glutamic acid but not when it is ammonium sulphate, although the host grows equally well on both. This effect could be due to either the direct toxicity of ammonium sulphate to the biotroph or to indirect effects brought about by the influence of the ammonium salt on host physiology. As with contact biotrophs, nutritional studies on dual systems are always open to a number of interpretations and it is experiments using axenic cultures that have so far contributed most to an understanding of the physiology of haustorial biotrophs.

It has been known for some time that it is possible to grow some haustorial biotrophs on natural media that are rich in protein, but it has not always been possible to grow these same species on chemically-defined media (Ayers, 1933; Ellis, 1966). Although few fungi have been studied using defined media, there is some evidence that all biotrophs may share a number of common physiological attributes (Kurtzman, 1968; Barnett, 1970; Barker and Barnett, 1973). None of the species so far studied has an absolute requirement for mycotrophein-like compounds for growth in axenic culture. *Dispira cornuta, Dispira simplex, Dimargaris verticillata, Dimargaris bacillispora* and *Tieghemiomyces parasiticus* do not readily utilize glucose and most species will probably not utilize a wide range of other mono- and disaccharides. Only one species, *Dispira parvispora*, has been found to readily utilize glucose (Barnett and Binder, 1973). However, good growth of most species will take place with glycerol as sole carbon source, and in the case of *Dispira cornuta* there is also good growth on acetate. An inability to grow on glucose is uncommon in fungi and in haustorial biotrophs this could be due to either lack of glucose movement into them or to a deficiency in those enzyme systems necessary for its utilization. However, in at least one species. *Tieghemiomyces parasiticus*, all EMP and HMP enzymes have been found to be present in cell-free mycelial extracts. Liberation of labelled carbon dioxide by this fungus when it is grown on either ^{14}C-labelled glycerol or glucose indicates that glucose is metabolized but at a much lower rate than glycerol. Available evidence, therefore, points to the existence of permeability barriers to glucose, and perhaps to other soluble carbohydrates as well, rather than to a metabolic deficiency. It has been suggested that glycerol acts as a surfactant and so facilitates entry of nutrients into hyphae, but this seems unlikely. Ability to utilize glycerol or acetate may be related to the usually lipid-rich nature of their host's mycelium.

Another common feature shown when these fungi are grown in axenic culture with glycerol as sole carbon source is a requirement for high nitrogen levels, a preference being shown for casein hydrolysate or yeast extract (Barnett, 1970; Barker and Barnett, 1973). With acetate as carbon source their nitrogen demands seem to be more modest (Kurtzman, 1968). With respect to requirements for specific nitrogen compounds, there are wide differences between different species. *T. parasiticus* will not grow with any single amino acid as sole source of nitrogen but requires a mixture of cysteine, leucine and valine (Binder and Barnett, 1973). In contrast, *Dispira cornuta* grows well on either asparagine, alanine, aspartic or glutamic acid, and is even able to utilize ammonium sulphate or urea, although there is no growth on nitrate nitrogen. Haustorial biotrophs also have a requirement for high levels of thiamine and may be deficient for biotin. However, one isolate of *Dispira cornuta* has been found to be autotrophic for both of these vitamins.

One species, *Spinellus macrocarpus*, stands apart from both contact and haustorial biotrophs. This fungus is a member of the Mucorales but is not merosporangial and its host range, unlike that of other biotrophs in the Mucorales, is restricted to Basidiomycotina fruit bodies, particularly those of *Mycena* species. Whether it is a contact or haustorial fungus is not clear, but on agar media its spores require basidiocarp extracts to stimulate germination (Ellis and Hesseltine, 1962; Watson, 1962; Leadbeater and Richardson, 1963; Watson, 1964). The active compound may be ascorbic acid. In axenic culture it can utilize glucose, fructose and even mannitol and trehalose as sole carbon sources and it may be significant that basidiocarps frequently contain considerable amounts of mannitol and trehalose (Watson, 1965*b*). Single amino acids do not provide suitable nitrogen sources for growth and *S. macrocarpus* is heterotrophic for both thiamine and nicotinic acid. This species therefore seems to share some physiological characteristics of both contact and haustorial biotrophs.

It is difficult to account for the apparent restriction of most haustorial fungi to hosts within the Mucorales. For example, the host range of *Dispira cornuta* has been studied in some detail, but of the wide range of fungi from all the major groups tested only species of the Mucorales can act as hosts (Mandelbrot and Erb, 1972). There are indications that only species of the Mucorales can supply a metabolite, so far not identified, which this and similar biotrophs require, and which stimulates their growth in axenic culture.

As with the necrotrophs it is not possible to assess the significance of biotrophic symbioses in natural ecological situations. Dual culture experiments suggest that, even if such associations were very common in nature, there would often be little or no effect on host growth. Experiments in dual culture using nutrient-rich media do not, however, give a good indication of host response under those nutrient conditions obtaining in nature, especially when the host may be under competitive stress from other micro-organisms.

Bibliography

Ainsworth, G. C. and Bisby, G. R., 1971. *Dictionary of the Fungi*, Commonwealth Mycological Institute; Kew.
Armentrout, V. N. and Wilson, C. L., 1969. *Phytopathology*, 59: 897—905.
Ayers, T. T., 1933. *Mycologia*, 25: 333—341.
Ayers, T. T., 1935. *Mycologia*, 27: 235—261.
Barker, S. M. and Barnett, H. L., 1973. *Mycologia*, 65: 21—27.
Barnett, H. L., 1958. *Mycologia*, 50: 497—500.
Barnett, H. L., 1963. *A. Rev. Microbiol.*, 17: 1—14.
Barnett, H. L., 1964. *Mycologia*, 56: 1—19.
Barnett, H. L., 1968. *Mycologia*, 60: 244—251.
Barnett, H. L., 1970. *Mycologia*, 62: 750—761.
Barnett, H. L. and Binder, F. L., 1973. *A. Rev. Phytopath.*, 11: 273—292.
Barnett, H. L. and Lilly, V. G., 1962. *Mycologia*, 54: 72—77.
Bean, G. A., 1968. *Phytopathology*, 58: 252—253.
Benjamin, R. K., 1959. *Aliso.*, 4: 321—433.
Berry, C. R., 1959. *Mycologia*, 51: 824—832.
Berry, C. R. and Barnett, H. L., 1957. *Mycologia*, 49: 374—386.
Binder, F. L. and Barnett, H. L., 1973. *Mycologia*, 65: 999—1006.
Boosalis, M. G., 1964. *A. Rev. Phytopath.*, 2: 363—376.
Brunk, M. A. and Barnett, H. L., 1966. *Mycologia*, 58: 518—523.
Buller, A. H. R., 1922. *Researches on Fungi*, Vol. 2, Longmans, Green, London.
Buller, A. H. R., 1924. *Researches on Fungi*, Vol. 3, Longmans, Green, London.

270

Butler, E. E., 1957. *Mycologia,* 49: 354—373.
Butler, E. E. and McCain, A. H., 1968. *Mycologia,* 60: 955—959.
Calderone, R. A. and Barnett, H. L., 1972. *Mycologia,* 64: 153—160.
Dayal, R. and Barron, G. L., 1970. *Mycologia,* 62: 826—830.
Dennis, C. and Webster, J., 1971. *Trans. Br. mycol. Soc.,* 57: 363—369.
De Vay, J. E., 1956. *A. Rev. Microbiol.,* 10: 115—140.
Dobbs, C. G. and English, M. P., 1954. *Trans. Br. mycol. Soc.,* 37: 375—389.
Ellis, J. J., 1966. *Mycologia,* 58: 465—469.
Ellis, J. J. and Hesseltine, C. W., 1962. *Nature, Lond.,* 193: 699.
Emmons, C. W., 1930. *Bull. Torrey bot. Club,* 57: 421—441.
England, W. H., 1969. *Mycologia,* 61: 586—592.
Gain, R. E. and Barnett, H. L., 1970. *Mycologia,* 62: 1122—1129.
Griffith, N. T. and Barnett, H. L., 1967. *Mycologia,* 59: 149—154.
Haskins, R. H., 1963. *Can. J. Microbiol.,* 9: 451—457.
Held, A. A., 1973a. *Can. J. Bot.,* 51: 1825—1835.
Held, A. A., 1973b. *Bull. Torrey bot. Club,* 100: 203—216.
Hunter, W. E. and Butler, E. E., 1975. *Mycologia,* 67: 863—872.
Indoh, H., 1965. *Trans. Mycol. Soc. Jap.,* 3: 60—64.
Karling, J. S., 1942. *Am. J. Bot.,* 29: 24—35.
Karling, J. S., 1960. *Bull. Torrey bot. Club,* 87: 326—336.
Kurtzman, C. P., 1968. *Mycologia,* 60: 915—923.
Leadbeater, G. and Richardson, M., 1963. *Nature, Lond.,* 198: 1015.
Linnemann, G., 1968. *Arch. Mikrobiol.,* 60: 59—75.
Mandelbrot, A. K. and Erb, K., 1972. *Mycologia,* 64: 1124—1129.
Manocha, M. S., 1972. *Can. J. Bot.,* 50: 35—57.
Manocha, M. S. and Lee, K. Y., 1971. *Can. J. Bot.,* 49: 1677—1681.
Nicolas, G. and Villanueva, J. R., 1965. *Mycologia,* 57: 782—788.
Rakuidhyasastra, V. and Butler, E. E., 1973. *Mycologia,* 65: 580—593.
Rambo, G. W. and Bean, G. A., 1970. *Phytopathology,* 60: 1436—1440.
Shigo, A. L., 1960. *Mycologia,* 52: 584—598.
Shigo, A. L., Anderson, C. D. and Barnett, H. L., 1961. *Phytopathology,* 51: 616—620.
Slifkin, M. K., 1961. *Mycologia,* 53: 183—193.
Swendsrud, D. P. and Calpouzos, L., 1970. *Phytopathology,* 60: 1445—1447.
Tubaki, K., 1955. *Nagaoa,* 5: 11—40.
Watson, P., 1955. *Trans. Br. mycol. Soc.,* 38: 409—414.
Watson, P., 1962. *Nature, Lond.,* 195: 1018.
Watson, P., 1964. *Trans. Br. mycol. Soc.,* 47: 239—245.
Watson, P., 1965a. *Trans. Br. mycol. Soc.,* 48: 9—17.
Watson, P., 1965b. *Trans. Br. mycol. Soc.,* 48: 73—80.
Whaley, J. W. and Barnett, H. L., 1963. *Mycologia,* 55: 199—210.
Willoughby, L. G., 1956. *Trans. Br. mycol. Soc.,* 39: 125—141.

Index

278

Musca autumnalis, 61
Mutualism, 4
Mutualistic symbionts, 10
Mutualistic symbioses, with algae,
    226—252
  with higher plants, 185—223
  with insects, 99—124
  with other fungi, 257, 265
Mutualistic symbiosis, definition, 4
Mycelium radicis, in ericaceous
    mycorrhiza, 206
  callunae, 206
Mycena, 268
Mycetangia, of Lymexylidae, 105
  of Platypodidae, 103
  of Scolytidae, 102—103
  of Siricidae, 106—109
Mycetocytes, 120—121
Mycetoma fungi, 69
Mycetomas, 69, 70, 95
Mycetome fungi, 119—124
  axenic culture, 123
  effect on host growth, 123—124
  effect on host's sex, 124
  location and transmission, 120—123
Mycetomes, 119—124
  in Anobiidae, 120—122
  in aphids, 122
  in Cerambycidae, 122
  in Cicadinae, 122
  infection by *Torulopsis utilis*, 124
Mycobiont, 227
Mycoparasitism, 257
Mycophycobioses, 226
Mycorrhizal fungi, *see* Ectomycorrhizal
    fungi and Endomycorrhizal fungi
Mycorrhizas, ectendo-, 185—186
Mycorrhizas, ecto-, 185—200
  auxins and, 198—200
  carbohydrate physiology, 188—192
  effect on host, 192—196
  formation, 186—188, 196—200
  fungi, 187
  mineral uptake, 192—196
  relationship with *Monotropa*,
    212—213
Mycorrhizas, endo-, 185—186, 188,
    203—223
  ericaceous, 204—214
  vesicular-arbuscular, 214—223
Mycorrhizas, morphological groups of,
    185—186
  orchidaceous, 130, 211
Mycoses, 70

Mycosphaerella, 170
  ascophylli, 226
Mycotrophein, 264
Mycotrophy, 213
Myrmecia, 232, 240
Myrmicine ants, 110—113
Myxomycota, 11, 136

Nannizia persicolor, 75
Necrotrophs, destroying nematodes,
    21—36, 43—51
  destroying protozoa, 15—18, 38
  obligate, 6, 129, 169—175
  on higher plants, 129—134, 169—175
  on insects, 53—64
  on other fungi, 258—261
  on vertebrates, 68—77
  trapping rotifers, 18—21
Necrotrophy, definition, 5
  in lichen fungi, 231, 251
Nectria coccinea var. *faginata*, 262
Nematoctonus, 45—51
  concurrens, 48, 49
  haptocladus, 46, 48
  tripolitanius, 50
Nematode-destroying fungi, 21—36,
    43—51
  axenic culture, 30—35, 47—48
  biological control by, 35—36, 51
  in soil, 32—35, 51
  morphogenesis, 30—31
  nutrient requirements, 32—35,
    47—48
  toxins, 29—30, 48—51
  trapping mechanisms, 21—28
  ultrastructure, 21, 23, 27—28
Nemin, 31
Neoaplectana glaseri, 31
Neurospora, 72
Nitrogenase, 245—246
Nitrogen fixation, 244—246
Nostoc, 227, 232, 241
  carbohydrate release from, 235—243
  nitrogen fixation in, 244—246
  muscorum, 235—236, 244, 245
Neutralism, 4
Neutral fungi, 83—95, 203
Neutral symbioses, 4—5, 8—9, 9—10
Nucleic acid metabolism, 146—151
  inhibition by cordycepin, 63
  in *Plasmodiophora* infections,
    147—151
  in rust fungi, 147
  in *Ustilago* infections, 151